Chaotic Dynamics of Fractional Discrete Time Systems

Vignesh Dhakshinamoorthy
Department of Basic Sciences and Humanities, School of Engineering and Technology
CMR University, Chagalahatti, Bangalore-562149

Guo-Cheng Wu
Data Recovery Key Laboratory of Sichuan Province
Neijiang Normal University, Neijiang, Sichuan, People's Republic of China

Santo Banerjee
Dipartimento di Scienze Matematiche
Politecnico di Torino, Corso Duca degli Abruzzi, Torino, Italy

CRC Press
Taylor & Francis Group
Boca Raton London New York

CRC Press is an imprint of the
Taylor & Francis Group, an **informa** business
A SCIENCE PUBLISHERS BOOK

First edition published 2025
by CRC Press
2385 NW Executive Center Drive, Suite 320, Boca Raton FL 33431

and by CRC Press
4 Park Square, Milton Park, Abingdon, Oxon, OX14 4RN

CRC Press is an imprint of Taylor & Francis Group, LLC

Library of Congress Cataloging-in-Publication Data (applied for)

ISBN: 978-1-032-54476-2 (hbk)
ISBN: 978-1-032-54481-6 (pbk)
ISBN: 978-1-003-42511-3 (ebk)

DOI: 10.1201/9781003425113

Typeset in Times New Roman
by Prime Publishing Services

Preface

In the complex network of dynamical systems, an intriguing and occasionally enigmatic subset emerges, disrupting established notions of continuity and order. This book explores the intriguing domain of chaotic dynamics within discrete fractional systems, a frontier where the amalgamation of fractional calculus and discrete dynamics creates a complex and rich terrain of unpredictability. Standing at the crossroads of fractional calculus and chaos theory, we find ourselves with a distinctive chance to investigate the subtleties of systems displaying both fractional order and chaotic tendencies.

The exploration of fractional calculus has offered a potent tool for examining phenomena characterized by non-integer order derivatives, enhancing our comprehension of intricate dynamics across diverse fields. Simultaneously, the evolution of nonlinear dynamics, commonly known as chaos theory, has transformed into a captivating discipline that surpasses traditional boundaries, influencing various realms of science and engineering. From the rhythmic flutter of a butterfly's wings to the intricate dynamics of financial markets, nonlinear systems pervade our surroundings, challenging established notions and reshaping our perspectives on predictability and determinism. This book aims to integrate these two realms, delving into the chaotic dynamics that emerge when fractional calculus intersects with discrete systems.

As we navigate the contents of this book, we are set to undertake an exploration delving into the fundamental principles of both fractional calculus and chaos theory. This expedition will intricately weave a narrative, revealing the inherent beauty and complexity encapsulated within discrete fractional systems. The path we traverse will not only unfold the sophisticated mathematics that forms the bedrock of these systems but will also extend to a exploration of real-world applications spanning various disciplines. Throughout this intellectual voyage, we will delve into the captivating interplay between fractional dynamics and chaos, unravelling the intricate connections and shedding light on the fascinating relationships that exist within these domains.

This book is crafted for scholars, researchers, and students eager to explore the uncharted territories of discrete fractional dynamics. Through a balanced blend of theoretical discussions, numerical simulations, and practical applications, it aims to provide a comprehensive resource for those seeking to understand and navigate the chaotic landscapes that emerge in systems characterized by both fractional order and discreteness.

As we embark on this journey, I express my gratitude to the pioneers in both fractional calculus and chaos theory who have paved the way. Their invaluable contributions form the bedrock upon which we aim to build insights and revelations within these pages. May this book stand not only as a guide but also as an inspiration for those captivated by the chaotic dynamics of discrete fractional systems, igniting curiosity and nurturing a profound appreciation for the intricacies that unfold when fractional calculus converges with the realm of chaos. Thank you for joining on this literary journey. Your curiosity and active participation hold immense value in my eyes.

Contents

Chapter 1

Mathematical Perspective of Real World Models

One of the most important and common interests in the science fraternity is to know how the world works. This requires a powerful tool to describe nonlinear phenomena accurately. However, conducting experiments to gain accurate insights can be expensive and prone to errors in construction, leading to significant time and financial investments for research groups. To address these challenges, the scientific community has embraced mathematical models as a powerful tool for depicting situations with precision while adhering to the laws of nature. The field of mathematical model construction has flourished, attracting interdisciplinary researchers who appreciate its ability to provide a multidimensional perspective on the phenomena under investigation. This chapter aims to enhance our comprehension of mathematical models by offering a comprehensive examination of their applications in various fields of study. Additionally, it delves into the fundamentals of fractional calculus theory and highlights the importance of studying fractional derivatives. Furthermore, a brief introduction to the development of discrete fractional calculus theory is provided, along with relevant mathematical prerequisites.

1.1 Mathematical Modelling

Mathematical modeling has been a prominent and significant field for several centuries, playing a crucial role in understanding real-life events and their behavior. Essentially, mathematical modeling serves as an essential tool for expressing real-world situations in a mathematical manner, incorporating relevant conditions. The field has experienced exponential growth since the advent of supercomputers and the increased computational power to tackle complex and intricate problems. By transforming real-world problems into mathematical forms, it becomes possible to find solutions and explore them numerically. For instance, determining the height of a tower using trigonometric principles based on angles and horizontal distances is a simple example of mathematical modeling. The primary challenge in developing mathematical models lies in incorporating all the factors involved in the functioning of real-world phenomena. However, continuous improvement of the model can be achieved by initially incorporating the most important and influential factors and subsequently exploring the less significant or less influential factors. The implementation of mathematical methods for analyzing a system may vary depending on small changes that occur within the system. Mathematical models must meet certain fundamental requirements, which include:

1. Adaptability to various situations,
2. Suitability or relevance to the events being considered,
3. Precise description of the characteristics of the event,
4. Cost-effectiveness.

The primary objectives of constructing mathematical models are to provide a clear scientific understanding through quantitative analysis, explore changes occurring in the system, and make important decisions based on system analysis. Mathematical models can be either deterministic or stochastic. Deterministic models predict future outcomes based on initial conditions, while stochastic models incorporate random variations and statistical data obtained from experiments, surveys, or real-life observations. Examples of deterministic models include pricing structures and economic models, while stochastic models, such as financial models, guide decision-making in investments. The process of modeling consists of several stages as follows:

1. Identification of a real life situation,
2. Model construction,
3. Mathematical analysis (finding solutions),
4. Interpretation of the consequences of the model,
5. Validation using real time data or results obtained through experiments to demonstrate the feasibility of the model.

An example of mathematical modelling using differential equations is discussed below.

Cerberospinal Fluid Model:
The significance of mathematical biology can be established from the model developed to examine the pressure of cerebrospinal fluid, which is a first-order differential equation of the form

$$Q(U)\frac{dU}{dt} + \frac{U(t)}{W_a} = J_f(t) + \frac{U_e}{W_a}, \; U(0) = U_0 \tag{1.1}$$

where U represents pressure, W_a represents fluid resistance, and U_e represents fluid's threshold pressure. Here, the function $Q(U)$ is essential to the investigation of the dynamical behaviour of the cerebrospinal fluid and is dependent on the volumetric change of the fluid with respect to pressure $\left(\frac{dV}{dU}\right)$. The rate of change in the volume of fluid can be determined using

$$\frac{dV}{dt} = J_f(t) - J_a(t). \tag{1.2}$$

Equation (1.2) is the difference between the rate at which cerebrospinal fluid is formed and the rate at which the fluid is absorbed. The formation of the fluid can be either natural or artificial, and the rate of absorption is dependent on the fluid's resistance and its threshold pressure, which can be expressed mathematically as follows:

$$J_a(t) = \frac{U(t) - U_e}{W_a}. \tag{1.3}$$

The rate of change of cerebrospinal fluid is highly dependent on the factors involved in fluid formation and how they interact with the various situations that may arise. To effectively analyse the hydrodynamics of the system, the function $Q(U)$ adopts distinct values. The simplest option is $Q(U) = K$, which corresponds to a constant rate of formation. However, the formation may not always be constant, which leads to the exponential or hyperbolic forms: $Q(U) = \frac{1}{KU}$ or hyperbolic form $Q(U) = \frac{a}{1 + bU}$. The function can be selected based on the experimental results of the system. The model can be expanded by introducing the function $S(t)$ to represent fluid recirculation diversion.

During the analysis phase of modeling, researchers employ various approaches, with qualitative and quantitative theories being the most common ones. Qualitative results are derived through rigorous mathematical analysis, and these results can then be compared with experimental data for validation. On the other hand, quantitative analysis involves investigating the system using comprehensive information and real-time datasets, often supported by simulations. In this case, the emphasis is less on the mathematical analysis and more on utilizing the available data for analysis purposes.

Discrete–Time Models:
Differential equations are commonly used to model phenomena where the system's states evolve smoothly over continuous time. In contrast, discrete models represent the state of an event or phenomenon at a countable number of discrete–time points. While continuous and discrete–time models share certain aspects of modeling, it would be incorrect to consider them equivalent. Instead, the choice of modeling method depends on the specific requirements of the situation, and there are several methods available for obtaining a discrete model from a continuous one, and vice versa. One significant advantage of employing discrete–time models is that they allow for more precise representation and analysis of systems that inherently operate in a discrete manner. Discrete models can accurately capture phenomena that exhibit distinct events or changes at specific time intervals, such as computer simulations, digital signal processing, or discrete event systems. By discretizing the time domain, these models provide an appropriate framework for studying and understanding such systems. Additionally, discrete–time models often lend themselves well to computational simulations and numerical analysis, as they naturally align with the discrete nature of computers and algorithms. This enables researchers to study and simulate complex systems more efficiently, taking advantage of the computational power and tools available.

1. Data sets representing real-world events are often obtained in discrete form, comprising specific points or intervals in time.
2. Computers operate using discrete units, which makes them suitable for simulating and analyzing discrete models.
3. Discrete models serve as a general framework, where the limit of infinitesimally small time intervals approaching zero results in a continuous-time system. In other words, continuous-time models can be seen as a special case of discrete-time models when the time intervals become infinitesimally small.

The following sections will delve into the significance of mathematical model building in various intriguing areas of research.

1.1.1 Biological Models

Biological systems, characterized by their complexity and nonlinear interactions, present significant challenges when it comes to modeling. The exchange of information between elements within these systems, often observed in population models, adds further complexity in accurately describing their behavior. While a quantitative approach can provide valuable insights into studying biological systems, the need to consider various factors complicates the task of defining system boundaries. Mathematical biology, which involves mathematically representing the workings and functions of living organisms, has become a vast and interdisciplinary field attracting biologists, ecologists, medical professionals, and others. Mathematical models offer a clear pathway to understanding biological phenomena. Mathematical models in biology encompass more than just the study of human organs and their functions; they also cover areas such as genomics, neural biology, cancer models, enzyme kinetics, population dynamics, physiological systems, and biophysics. The growth of this field can be attributed to its ability to provide accurate mathematical representations of complex biological phenomena and describe the behavior of nonlinear systems in a meaningful way.

1.1.1.1 Population Dynamics

Population dynamics is a vital field within mathematical biology, focusing on modeling the growth of populations, as well as the survival and coexistence of different species within specific ecosystems. Over the past century, this field has experienced significant expansion, with the construction of numerous new models, making it one of the most extensively explored sub-fields of mathematical biology. The origins of population dynamics can be traced back to Thomas Malthus, who studied population growth models. In the 19th century, researchers such as Gompertz and Verhulst further advanced demographic studies by modifying the Malthusian model. However, it was the groundbreaking work of Lotka and Volterra in the early 20th century on prey-predator interactions that revolutionized the field of population dynamics. Lotka initially proposed equations to represent autocatalytic reactions in 1910, but it was Volterra who, in 1926, utilized the same set of equations to describe species interactions.

Prey–Predator Interaction Model:
The primary purpose of constructing an ecological model is to investigate the potential coexistence of different species within an ecosystem and prevent species extinction. A simple mathematical model can be developed by considering the prey population at time n as $p(n)$ and the predator population at time n as $q(n)$. However, to develop such a model, it is necessary to make certain assumptions about the ecosystem and environmental factors. The first basic assumption involves the availability of food for the prey population and assumes exponential growth of the population. The model typically neglects any environmental bias towards either population. Additionally, the model assumes that the predator population's intake of food is unrestricted and directly proportional to the available prey population. It also assumes that the predator does not switch resources in the absence of prey. Based on these assumptions, the prey population is expected to grow exponentially at a rate δp, while the decrease in the population can largely be attributed to predation by the predator population, represented by $\eta p\, q$. Consequently, the change in the prey population over time can be expressed using a mathematical formulation.

$$p_{n+1} = p_n \left[1 + \delta - \eta q_n\right]. \tag{1.4}$$

The second term in the prey population equation represents the decrease in prey population, denoted by a negative sign. Similarly, the growth of the predator population is described by the term $\beta p\, q$, where a different rate constant, denoted as β, is used. This is because the rate of predator growth does not solely depend on the food intake through hunting. The decrease in the predator population can be attributed to factors such as the absence of food, death, or movement away from the ecosystem, which is denoted by $d\, q$. Thus, we have

$$q_{n+1} = q_n \left[\beta p_n - d + 1\right]. \tag{1.5}$$

Combining equations (1.4) and (1.5) allows one to arrive at a system that accurately represents the populations of two species in an ecosystem:

$$\begin{cases} p_{n+1} = & p_n \left[1 + \delta - \eta q_n\right], \\ q_{n+1} = & q_n \left[\beta p_n - d + 1\right]. \end{cases} \tag{1.6}$$

It is possible to make an extension to the system by introducing the logistic growth rate rather of the exponential growth rate. This growth rate more accurately portrays the behaviour of the system in real time since it takes into consideration the limitations on population increase. Then the system (1.6) becomes,

$$\begin{cases} p_{n+1} = & p_n \left[1 + \delta \left(\dfrac{C - p_n}{C}\right)\right] - \eta p_n q_n, \\ q_{n+1} = & q_n \left[\beta p_n - d + 1\right], \end{cases} \tag{1.7}$$

with C denoting carrying capacity.

Functional Responses:

The understanding of species interactions has been significantly enhanced by the introduction of functional responses, which capture the predation behavior, and the Allee effect, which explores the relationship between population density and species survival. As illustrated in the prey-predator model mentioned earlier, linear interactions may not accurately represent the dynamics, particularly when prey population density is high or when accounting for time delays in predator hunting. In such cases, it is more appropriate to replace the linear interaction with a nonlinear function that considers the density of the prey population and the rate of predation. C.S. Holling was the first to introduce the concept of functional responses, and his classification of three different types is widely accepted and still employed in modeling practices today. Some widely accepted and realistic functional responses are given below

1. Holling Type- I Response:
   The type - I functional response is defined by the limit of prey consumption,

   $$V_I(p_n) = \min\{\eta p_n, Z\},$$

   where Z represents the rate of prey consumption when their availability is abundant.

2. Holling Type- II Response:
   In type-II responses, total time (T) is split into the amount of time necessary by the predator for seeking prey (T_s) and the amount of time required for the predator to handle the prey ($T_h = r\eta p_n T_s$). The fundamental assumption behind this functional response is that the density of the prey and the search rate (η) of the predator are independent of one another. The functional response takes the form

   $$V_{II}(p_1) = \frac{\eta\, p_n}{1 + r\eta\, p_n}.$$

3. Holling Type- II Response:
   This functional response is a generalisation of the type-II reaction based on predation. Some examples of this kind of response include moving to various preys, predation when the prey density is low, optimum predation, and so on,

   $$V_{III}(p_n) = \frac{\eta\, p_n^{\nu}}{1 + r\eta\, p_n^{\nu}},$$

   where $\nu \geq 1$, $\nu = 1$ implies a type - II response.

4. Beddington–DeAngelis Functional Response
   Functional responses of the Holling type focused primarily on the population density and the rate of predation, but the interference of predators and their population density is taken into consideration while

developing the following functional response.

$$V(p_n, q_n) = \frac{\eta p_n}{1 + \eta\, r\, p_n + \ell\, q_n},$$

where ℓ denotes predator interference.

Numerous mathematicians are now engaged in on-going research aimed at gaining a deeper comprehension of the system dynamics with a variety of functional responses. The study of the ecological system has also been improved by the incorporation of social effects among the species, such as the fear factor, harvesting, the influence of the wind, cannibalism, and other such phenomena.

1.1.1.2 Epidemic Models

Epidemic models can be seen as an extension of population dynamics, offering mathematical insights into the transmission of diseases within the human population. This sub-field has garnered significant interest from medical professionals who seek a better understanding of the spread of diseases and the ability to predict infections over time. One crucial aspect of epidemiology is assessing the outcome of disease transmission in relation to populations at a risk of infection. Analyzing high-risk population groups is of utmost importance in order to prevent a major catastrophe. This is where the mathematical modeling of the situation becomes particularly valuable and appealing.

SIR Model:
SIR models serve as the foundation for studying epidemic and pandemic situations. These models involve a simple classification of a fixed population (N) into three compartments: the susceptible compartment (S), which includes individuals who are susceptible to the illness; the infected compartment (I), which includes individuals who are currently infected; and the recovered compartment (R), which includes individuals who have recovered from the infection. It is important to note that $S + I + R = N$, implying that the total population remains constant. Similar to prey-predator models, certain assumptions are necessary when constructing a SIR model. One such assumption is that no individuals are added to the susceptible compartment, and they are only removed from it only if they become infected with the illness. Additionally, the recovery rate of infected individuals is assumed to be constant. The equation representing the susceptible compartment contains only a single term since the only possibility, according to our assumptions, is removal by infection. Individuals become infected if they come into contact with someone who is already infected. Therefore, the equation representing the susceptible population takes the following form.

$$S_{n+1} = S_n - \frac{\eta S_n I_n}{N}, \tag{1.8}$$

where η is the rate of contact between susceptible and infected persons.

The equation for the infected compartment can be derived as an extension of the susceptible compartment equation. As we can observe, the increase in the population in this compartment depends on the rate at which the susceptible population becomes infected, and the decrease in the population is determined by the recovery rate (δ) from the infection. Then the equation takes the form

$$I_{n+1} = \frac{\eta S_n I_n}{N} - (\delta - 1)\, I_n, \tag{1.9}$$

Finally, the recovery compartment can be expressed as

$$R_{n+1} = R_n + \delta\, I. \tag{1.10}$$

Combining equations (1.8), (1.9), (1.10), we get

$$\begin{cases} S_{n+1} = & S_n - \dfrac{\eta S_n\, I_n}{N}, \\ I_{n+1} = & \dfrac{\eta S_n\, I_n}{N} - (\delta - 1)\, I_n, \\ R_{n+1} = & R_n + \delta\, I. \end{cases} \tag{1.11}$$

The system given by (1.11) represents a basic form of an epidemic model. However, this system can be expanded by incorporating additional compartments to capture more realistic situations. For example, a death compartment can be included to study the dynamics of the model with a mortality factor. A quarantine compartment can be introduced to account for isolation measures. The infection compartment itself can be divided into two compartments to represent symptomatic and asymptomatic infections. Other compartments, such as those for hospitalization and vaccination, can also be included. Analyzing the system can be done by considering practical and relevant scenarios, such as reinfection, environmental influences, immunity, co-morbidity, and fear factor. The success of epidemic models relies on including influential factors that are crucial for the spread of the disease. Neglecting these factors may result in the system exhibiting weaker dynamic behavior compared to real-time data. The significance of studying epidemic models is particularly evident during the COVID-19 pandemic, where precautionary measures and predictions have played a vital role in preventing mass fatalities to a certain extent.

1.1.1.3 Neuron Models

In the twentieth century, there has been a growing interest among humans to understand the nature of the brain and the nervous system. Extensive research has been conducted to comprehend how people think and react to external stimuli and internal messages transmitted through neurons. The development of artificial intelligence and machine learning techniques has heavily relied on the mathematical study of neural networks. However, comprehending the functioning of the nervous system is a complex task due to its intricate structural components. Nevertheless, it is possible to mathematically construct and understand the nervous system by incorporating physiological principles of neural functions. Modeling neurons and gaining a thorough understanding of their intricacies is a challenging endeavor, as it requires careful consideration of numerous parameters.

Hodgkin–Huxley Model of Neurons:
The Hodgkin-Huxley neuron model is one of the very few system representations of neurons that can produce both qualitative and accurate numerical predictions. This makes it one of the most useful models available. In this model, the potential of the neuron is accorded the utmost significance, and the research is carried out on the basis of the differential equation shown in the following expression:

$$C\frac{dP}{dt} = K_e - K_i, \tag{1.12}$$

where the membrane capacitance of the neuron is represented by C and the internal current and exterior current are respectively denoted by K_i and K_e. Here, the internal current is expressed by the equation,

$$K_i = h_0(P - P_0) + h_1 l^3 g(P - P_1) + h_2 r^4 (P - P_2), \tag{1.13}$$

where $h_0, h_1, h_2 > 0$ and P_0, P_1, P_2 are constants, while the inflow of Na^+ ions through the opening of sodium channels is denoted by l, outflow of K^+ ions through the opening of potassium channels and closing of sodium ion channels are represented by g and r respectively. The increasing focus on constructing models of neurons under various realistic scenarios aims to estimate the physical properties and important characteristics related to information processing. Discrete–time models of neurons have also been extensively analyzed and given significant importance due to their computational feasibility, simplicity, and reliability in modeling.

Chialvo Model:
The following nonlinear discrete–time model shows how the activation variable (u_n) and the recovery variable (v_n) interact:

$$\begin{cases} u_{n+1} = & u_n^2 e^{v_n - u_n} + \kappa, \\ v_{n+1} = & \delta v_n - \eta u_n + \alpha, \end{cases} \tag{1.14}$$

where the parameter κ represents the time-dependent additive perturbation, while δ, η, and α play a crucial role in determining the fixed points for the recovery state variable. The Chialvo model captures the excitation of neurons in various biological systems, including nerves, pancreas, and the heart. Analyzing the behavior of these systems using discrete–time models proves valuable in understanding the complex chaotic dynamics exhibited by neurons. Discrete–time models provide insights into the spiking behavior of neurons and enhance our understanding of their intricate functioning.

1.1.1.4 Growth and Dynamics of Tumor Cells

Cancer, a prevalent and challenging disease in the 21st century, has attracted significant attention and resources from both biological and mathematical research communities. The increasing rate of cancer cases necessitates interdisciplinary studies involving experts from various fields such as biology, mathematics, immunology, and physics. The collaboration between these disciplines has led to significant advancements in understanding cancer through the analysis of mathematical models. The integration of mathematics and computer science has played a crucial role in visualizing and interpreting cancer-related phenomena. By combining mathematical models with computational tools, researchers are able to gain deeper insights and make better predictions regarding cancer dynamics. However, despite numerous scientific observations, new discoveries in the field have often added complexity to our understanding of cancer systems. Specific molecules such as cytokines, chemokines, interferons, and growth factors have been found to have diverse biological effects, with their roles in the immune system varying depending on the circumstances. Over the past few decades, mathematical models of tumor growth have been developed with the primary aim of providing qualitative descriptions of the different stages of tumor cell growth and assessing the stability of these cells.

Competition Model for Tumor Growth:
Mathematical models for immune response often employ the prey-predator approach, where the predator population represents lymphocytes produced by the body, and the prey population represents tumor cells. These models analyze the interaction between lymphocytes and tumor cells under different circumstances.

The growth of tumor cells is influenced not only by their own dynamics but also by the interaction with lymphocytes. Therefore, any mathematical model on tumor immune interaction should consider this interaction mechanism, regardless of the artificial therapies used to control tumor growth. It is crucial to account for the competition among different cell populations when studying the dynamics of tumor cells. The interaction between normal cells in the body and tumor cells can be mathematically represented as a simple model, where the changes in the tumor cell population are influenced by the interaction with normal cells.

$$\begin{cases} C_{n+1} = & g_1 C_n(1 - a_1 C_n) + C_n - \delta_1 C_n T_n, \\ T_{n+1} = & g_2 T_n(1 - a_2 T_n) + T_n - \delta_2 C_n T_n, \end{cases} \tag{1.15}$$

where C_n and T_n represents the normal and tumor cell populations in time n. The system (1.15) assumes that both the normal and tumor cell populations grow logistically at rates g_1 and g_2, respectively. The decrease in population due to interaction with the other cell types occurs at rates δ_1 and δ_2, respectively. This model represents a simple interaction between normal and tumor cells. To incorporate the impact of immune cells, a separate compartment can be introduced to represent the types of cells that occur naturally or are artificially induced through various therapies. There are different types of growth functions that can be included to investigate tumor dynamics effectively. Here are some examples of effective models for the tumor cell population, denoted by T_n, and their growth rates, denoted by μ:

1. Exponential growth model:
   The mathematical representation takes the following form:
   $$T_{n+1} = \mu T_n.$$
2. Logistic model:
   This is the most common model to describe the gowth of cells as described in the system (1.15). The general form of the logistic model with carrying capacity N is,
   $$T_{n+1} = T_n \left(1 + r\left(1 - \frac{T_n}{N}\right)\right).$$
3. Gompertz law:
   Gompertz law is represented as,
   $$T_{n+1} = T_n \left(\frac{T_n}{\beta}\right)^{\alpha \log k_0},$$
   where, β, α, k_0 are real constants.

The previously listed are a few of the growth models that are widely used to investigate tumour cells. Similar to this, a large number of therapeutic approaches are utilized in tumour systems to learn more about their dynamics. Some of these include direct cell death, antibody-induced cell death, DNA damage during cell cycle stages, immune cell introduction, and growth inhibition by interfering with cell phase cycles.

Cancer is a major cause of mortality for a significant portion of people in both developing and industrialised nations. One in five individuals get malignant tumour cells over their lifetime, and in 2020, there were almost 10 million documented fatalities. In 2020, there will be 83,300 cancer deaths in Canada and 606,520 cancer deaths in the USA. Researchers in several sectors are still working to develop effective cancer treatments that will only affect malignant cells and not harm healthy ones. Radiation therapy, surgery, chemotherapy, and immunotherapy are all treatment modalities that seem to be successful against cancer. Several key factors influence the development of cancer treatments. These factors include:

- Specific targeting: The treatment should effectively eliminate cancer cells while minimizing harm to healthy cells. It is crucial to selectively target cancer cells to minimize side effects and maximize the effectiveness of the treatment.
- Immune system enhancement: The treatment should be designed to enhance the immune system's ability to identify and destroy cancer cells. By strengthening the immune response against cancer, the treatment can improve the body's natural defense mechanisms.
- Immune system preservation: The treatment should not deplete the immune system to the extent that it compromises the body's ability to safeguard against other infections and diseases. It is important to maintain a balanced immune response to ensure overall health and well-being.

By considering these factors, researchers and healthcare professionals can develop treatments that are effective, targeted, and compatible with the body's immune system, ultimately improving patient outcomes in the fight against cancer.

1.1.2 Chemical Models

In general, two or more particles (referred to as reactants) are combined to create new particles (referred to as products) in a chemical reaction.

$$R_1 + R_2 \xrightarrow{q} P, \tag{1.16}$$

where R_1 and R_2 are the reactants and P is the product formed with rate constant q. Mathematical models of chemical reactions are simple transformations of physical information into mathematical equations by the laws of the mass action.

$$\frac{dP}{dt} = k[R_1][R_2]. \tag{1.17}$$

Modeling reactions allow for simulations that provide flexibility in exploring the behavior of reactions under different conditions. The mathematical framework is particularly necessary in chemical reactions where the time required to obtain a product can be lengthy. Model construction also provides insights into determining the appropriate amount of chemicals needed to achieve desired results. During a chemical reaction, the concentration of reactants decreases while the concentration of products increases until an equilibrium state is reached. The law of mass action governs most chemical reactions, where the equilibrium state is approached asymptotically through a simple reaction. However, some chemical reaction systems do not have a smooth approach to equilibrium and can exhibit prolonged oscillatory behavior. Chemical processes form the foundation of life, and numerous oscillatory processes have been identified in biological systems, such as insulin secretion, glycolytic reactions in yeast and muscles, and hormone cycles. The study of chemical reactions in biological species is vital for understanding the mechanisms underlying human and plant survival. An enzymatic reaction can be modeled as follows:

$$R_1 + R_2 \underset{q_{-1}}{\overset{q_1}{\rightleftharpoons}} Y \xrightarrow{q_2} P_1 + R_2, \tag{1.18}$$

where the substrate, enzyme, complex and product are represented by R_1, R_2, Y and P_1 respectively. The corresponding mathematical formulation can be given by system of four equations

$$\begin{cases} \frac{dR_1}{dt} &= q_{-1}[Y] - q_1[R_1][R_2] + F \\ \frac{dR_2}{dt} &= (q_{-1} + q_2)[Y] - q_1[R_1][R_2] \\ \frac{dY}{dt} &= q_1[R_1][R_2] - (q_{-1} + q_2)[Y] \\ \frac{dP}{dt} &= q_2[Y] - F_1 \end{cases} \tag{1.19}$$

where F and F_1 denote the rates of substrate supply and product removal, respectively.

Oscillating chemical reactions arise from a thermodynamic equilibrium state, typically involving at least one autocatalytic step and exhibiting two steady states under specific initial conditions. The study of spatial oscillations in biochemical systems has been a subject of extensive research for over three decades.

Chemical reaction systems are commonly developed with a focus on reaction rates. The periodic changes in the concentration of the chemical compounds involved form the basis of chemical oscillators. One well-known example of an oscillatory system is the Belousov-Zhabotinskii reaction, which serves as a classical example in non-equilibrium thermodynamics. These reactions often exhibit chaotic behavior and display various patterns, making them intriguing subjects of study. The discovery of the Belousov-Zhabotinskii reaction can be credited to Boris Belousov during his investigation of the Krebs cycle. Initially, Belousov's findings were met with skepticism due to the lack of a comprehensive explanation for the proposed theory. However, further research led by Anatol Zhabotinsky shed light on the reaction. The reaction itself is highly complex, involving 18 steps, and has been investigated by numerous researchers. The simplest model for the Belousov-Zhabotinskii reaction is the Oregonator model, which consists of five reactions.

$$\begin{aligned} R_1 + R_2 &\xrightarrow{q_1} U + P_1, \\ U + R_2 &\xrightarrow{q_2} 2P_1, \\ R_1 + U &\xrightarrow{q_3} 2U + 2V, \\ 2U &\xrightarrow{q_4} R_1 + P_1, \\ R_4 + V &\xrightarrow{q_5} \frac{1}{2} g\, R_2, \end{aligned} \tag{1.20}$$

where, $q_i, i = 1, 2, 3, 4, 5$ are the rate constants. The system representation is

$$\begin{cases} \frac{dU}{dt} &= q_1R_1R_2 - q_2U\,R_2 + q_3R_1\,U - 2q_4U^2 \\ \frac{dR_2}{dt} &= -q_1R_1R_2 - q_2U\,R_2 + \frac{1}{2}g\,q_5R_4\,V \\ \frac{dV}{dt} &= 2q_3R_1\,U - q_5R_4\,V. \end{cases} \tag{1.21}$$

In the case of chemical reaction models, discrete–time models are not constructed explicitly; rather, numerical methods such as the Euler approximation method, the Runge-Kutta method are applied to continuous time models in order to convert them to discrete–time models.

1.1.3 Physical Models

The mathematical representation of physical systems traces its origins back to the time of Sir Isaac Newton, who laid the foundation for modeling through the laws of physics. Newton's discovery of gravitation was a groundbreaking idea that enabled the study of celestial motion and planetary behavior within the solar system. This idea was later adopted by scientists to develop models and analyze the motion of objects, leading to advancements in various scientific fields.

In subsequent generations, scientists expanded upon Newton's concepts to study the evolution of physical systems using differential equations. Initially, understanding the behavior of systems described by differential equations was challenging, as there was a lack of knowledge regarding the solutions and difficulties in representing them algebraically with finite terms. The behavior of these systems was often characterized by steady states or periodic oscillations. However, during the latter part of the 20th century, the understanding of system behavior took a significant shift with the introduction of the concept of chaos—a phenomenon characterized by unpredictable and random motion. This expanded the viewpoint on the behavior of physical systems and brought forth a new dimension of analysis. The term "physical systems" in this context encompasses engineering systems such as mechanical, electrical, electromechanical, thermal, and hydraulic systems. Examples of physical systems include pendulum models, ion motion, suspension bridges, and models of electrical and electronic systems, among others.

Pantograph System:
The special form of delay differential equation describing the motion of a pantograph head is given by,

$$\phi'(\varpi) = a\phi(\varpi) + b\phi(\lambda\varpi), \varpi > 0, \tag{1.22}$$

where $a, b \in \mathbb{R}$ with $0 < \lambda < 1, \lambda \in \mathbb{R}$. Ockendon and Tayler discovered the pantograph equation while analysing the motion of the electric locomotive. The pantograph is introduced so that the pace of the locomotive or tram can be controlled. Engineers and mathematicians have examined the pantograph equation as part of their research. The equation is the modified version due to the modification of pantograph designs. Recent advances in science and technology have resulted in the conversion or fabrication of circuit models for physical systems so that they closely resemble real-time behaviour. Several system enhancements take the system's prior performance into account. The most significant improvement is the advancement of fractional calculus theory.

1.2 History of Fractional Calculus

Fractional calculus is a branch of mathematical analysis that deals with differentiation and integration operators of arbitrary order, extending beyond the classical integer orders used in ordinary calculus. Although the field gained attention in the 19th century, its origin can be traced back to the same period as ordinary calculus. The concept of non-integer order derivatives and anti-derivatives arose from a question posed by L'Hopital to Leibniz, who responded with the anticipation of useful consequences from this apparent paradox [1]. This idea sparked the interest of many renowned mathematicians, including Euler, Lagrange, Fourier, Laplace, Abel, Lacroix, Caputo, Riemann, Holmgren, Grunwald, Letnikov, Liouville, Weyl, and others [2, 3]. However, the simultaneous development of fractional calculus alongside classical calculus faced challenges in interpreting the physical and geometrical significance of non-integer order derivatives. Despite the efforts of these mathematical minds, it was Lacroix [4] who provided the first affirmative answer to the question on semi-derivatives posed by L'Hopital and Leibniz in 1819. Lacroix also developed a general formula for finding fractional derivatives of power functions by replacing factorial functions with gamma functions. In 1823, Abel [5] made significant progress by applying fractional order derivatives to the tautochrone problem, thereby establishing a physical application for these mathematical concepts.

Abel's work on fractional derivatives had a profound impact on the field, attracting the attention of Liouville, who conducted extensive research and made significant contributions to the theoretical aspects of fractional calculus based on classical calculus principles. Liouville's work led to the development of Riemann-Liouville type operators, which played a crucial role in the advancement of fractional calculus [6, 7, 8].

Following this development, several extensions of the definition were introduced by various mathematicians. A.V. Letnikov proposed an extension in [9], Laurent in [10], Grunwald and Letnikov in [11], O. Heaviside in [12], Weyl in [13], Hardy in [14], Marchaud in [15], Riesz in [16], Caputo in [17], and R. Hilfer in [18]. Among these contributions, Caputo's work was particularly significant as it incorporated classical initial conditions into the fractional calculus framework. Caputo replaced the initial conditions with non-integer order derivatives, which added physical significance to the models using Riemann-Liouville type derivatives. Recent contributions to fractional calculus include works of Wu et al. in [28, 29]. To provide a mathematical perspective on fractional calculus, the necessary definitions and theorems are presented below.

1.2.1 Why Fractional Calculus?

Traditionally, real-life events and physical phenomena have been described using differential equations with fixed integer orders. However, in recent years, there has been a shift towards using fractional order derivatives and integrals in modeling, surpassing the classical calculus approach. This shift is driven by the unique characteristics of fractional order operators, such as their non-local nature and ability to capture memory effects within the phenomena under consideration. The fractional order in differential equations serves as a parameter that allows for the exploration of system properties across a range of values. This introduces an additional dimension to the study of the system's dynamical behavior.

Lorentz System:

The following is a representation of the Lorenz equation, a mathematical model of atmospheric convection devised by Edward Lorenz.

$$\begin{cases} \dfrac{du}{dt} &= \sigma(v-u) \\ \dfrac{dv}{dt} &= u(\rho - w) - v \\ \dfrac{dw}{dt} &= uv - \beta w \end{cases} \tag{1.23}$$

where, σ, ρ, β are real parameters. In order to introduce memory effects into the system (1.23), we use the fractional derivative to replace the classical integer order derivative and obtain,

$$\begin{cases} \dfrac{d^{\vartheta}u}{dt^{\vartheta}} &= \sigma(v-u) \\ \dfrac{d^{\vartheta}v}{dt^{\vartheta}} &= u(\rho - w) - v \\ \dfrac{d^{\vartheta}w}{dt^{\vartheta}} &= uv - \beta w \end{cases} \tag{1.24}$$

where, $\frac{d^{\vartheta}}{dt^{\vartheta}}$ is the Caputo type fractional derivative with order ϑ. Because the fractional derivative operator has the feature of memory effects we use it to describe the nonlinear phenomenon with memory or non locality. To highlight the impact of a fractional order on the qualitative behavior of a system, the Lorentz system is considered as an example. The system is compared numerically with both an integer order model and a fractional order model. The fixed parameter values $\sigma = 14$, $\rho = 28$, and $\beta = 5$ are used, along with the initial condition $(u(0), v(0), w(0)) = (1, 1, 1)$. Numerical simulations are conducted for two cases: first, for the system (1.23) with an integer order model with $\vartheta = 1$, and second, for the system (1.24) with a fractional order model when $\vartheta = 0.9$. These simulations aim to provide a clearer illustration of the influence of fractional order on the qualitative behavior of the systems.

Figure 1.1 displays the phase plane analysis of the Lorentz system using integer order derivatives. The system exhibits chaotic behavior for the given parameter values, and the 3-D diagram along with corresponding 2-D phase plane diagrams are provided to illustrate this behavior. In Figure 1.2, the aperiodic oscillation of the state variables is demonstrated through time series analysis. A clear transition can be observed from chaotic to stable behavior when comparing Figures 1.1 and 1.2. This highlights the impact of fractional order on the qualitative behavior of the system, indicating the additional flexibility offered by constructing models with fractional order derivatives. Figure 1.4 showcases the varying amplitude of oscillations for different fractional orders in the system (1.24). Models constructed with fractional order derivatives exhibit improved accuracy compared to classical differential equations, particularly in relation to physical models. The non-locality aspect of fractional calculus is a key factor that captures the interest of scientists and engineers, as it allows for the incorporation of memory effects in various applications. Memory effects refer to the dependence of a system's behavior not only on its current state but also on its past states. Ordinary calculus is inadequate for

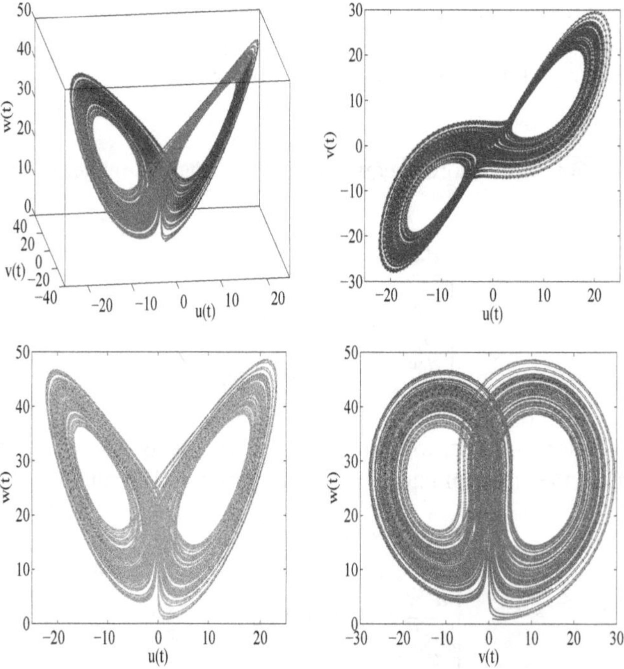

Fig. 1.1: Phase plane portraits for the integer order Lorentz system (1.23).

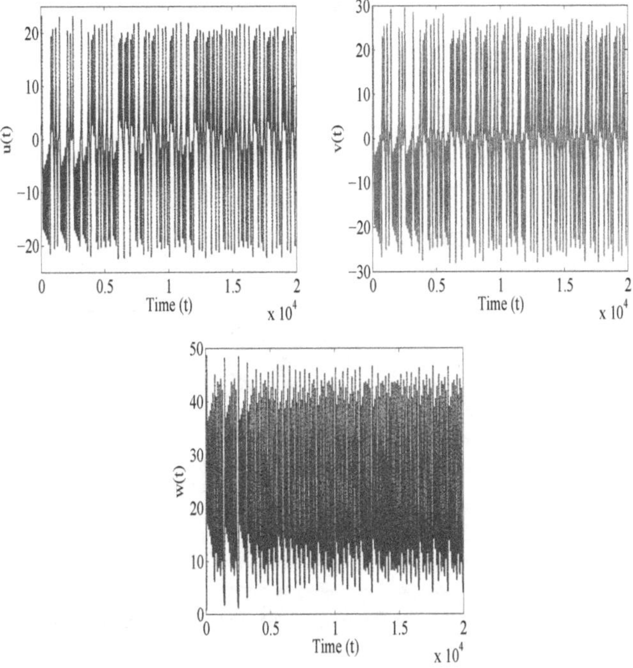

Fig. 1.2: Time series analysis for the integer order Lorentz system (1.23).

accurately modeling such memory-dependent systems, making fractional calculus a powerful tool for their analysis. The interpretation of fractional order derivatives in physical and geometrical terms remains an active research field in the 21st century, with widespread applications across science and technology, including viscoelasticity, electrical circuit theory, physical and chemical sciences, ecological systems, epidemic and pandemic modeling, neuron modeling, diffusion, tumor growth models, and more.

1.3 Discrete Fractional Calculus

There are many discrete-time models in real world application. However, it is difficult to introduce discrete memory effects into such models to describe long-term interactions of discrete states. However, the traditional numerical discretization causes losses in memory effects and there is no theoretical support. Fortunately, the time scale theory unifies the continuous and discrete-time systems. Discrete fractional calculus defined by the theory has nice propositions such as the semi-group property, summation by parts and the Leibniz summation law. So recently, the theory and applications have grown rapidly for analyzing and understanding systems that exhibit fractional behavior in a discrete setting. Discrete fractional calculus is an extended version of classical difference calculus and serves as the discrete counterpart to non-integer order differential equations discussed earlier. It deals with real-valued functions that involve shifts based on natural numbers. The field of discrete fractional calculus emerged in 1974 with the initial studies by Diaz and Osler. However, it initially gained limited popularity due to the lack of theoretical support. The development of the theory of discrete fractional calculus by Miller and Ross in 1989 [31], Podlubny in 2000 [19], Bohner et al. in 2001 [30], Atici and Eloe in 2007 [20], and Anastassiou in 2009 [21], Holm in 2011 [32] revived interest in the field,

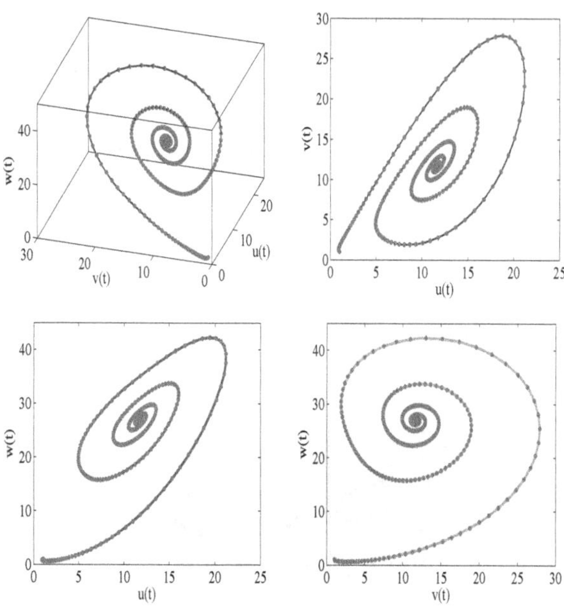

Fig. 1.3: Phase plane portraits for the fractional order Lorentz system (1.24) $\vartheta = 0.9$.

leading to significant growth and applications in various scientific and engineering domains. Both forward (Delta) and backward (Nabla) difference operators have seen parallel development in discrete fractional calculus. Notable researchers such as Christopher Goodrich [22], Abdeljawad and Baleanu [23], Wu [24], among others, have made significant contributions to the field. Discrete fractional systems hold particular importance for ecologists as they provide better descriptions of prey-predator interactions in ecosystems and epidemic models. The complex and chaotic dynamics exhibited by discrete fractional equations make them valuable for secure communications involving image encryption, image and voice processing, and the transmission of critical messages.

The field of discrete fractional calculus is relatively new and offers ample opportunities for exploring various essential features that can contribute to technological advancements and a better understanding of the real world. This makes it an intriguing area of study. Similar to continuous-time fractional operators, discrete fractional calculus defines different operators such as Caputo, Riemann-Liouville, and Hilfer types. However, the commonly used operators for modeling in this field are the Caputo and Riemann-Liouville operators. To provide a comprehensive mathematical understanding of the field, this section presents fundamental definitions and theorems. Some of the preliminary results related to Riemann-Liouville fractional operators are as under.

Definition 1.1 ($\beta - th$ Fractional Sum) For $\beta > 0$, the $\beta - th$ fractional sum of χ is

$$\Delta^{-\beta}\chi(\ell) = \frac{1}{\Gamma(\beta)} \sum_{s=i}^{\ell-\beta} (\ell - s - 1)^{(\beta-1)} \chi(s), \tag{1.25}$$

where χ is interpreted for $s = i\, mod(1)$ and $\Delta^{-\beta}\chi$ for $\ell = (i + \beta) mod(1)$ and $\Delta^{-\beta} : N_i \to N_{i+\beta}$.

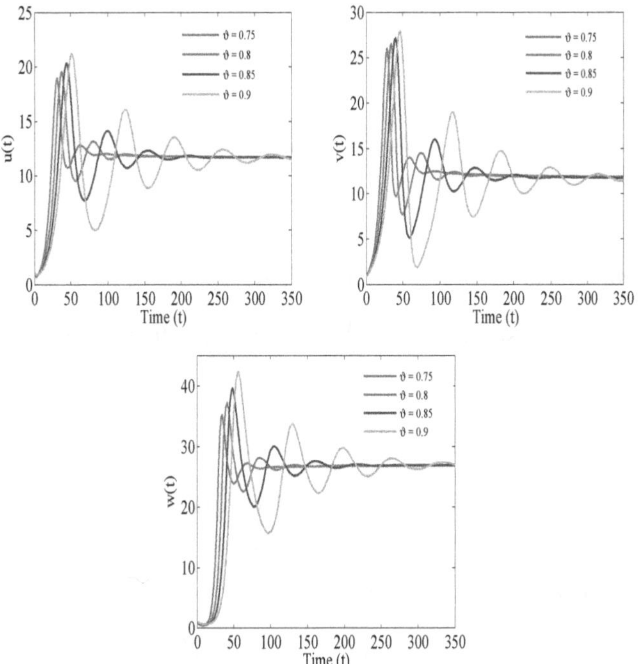

Fig. 1.4: Time series analysis for the fractional order Lorentz system (1.24).

Definition 1.2 Let λ be positive real number and $k-1<\lambda<k$, where $k \in \mathbb{Z}^{+}$, $k=\lceil \mu \rceil$, $\lceil . \rceil$ is the number ceiling. Assign $\beta = k-\lambda$. The $\lambda - th$ fractional difference is

$$\Delta^{\lambda}\chi(\ell) = \Delta^{k}\left(\Delta^{-\beta}\chi(\ell)\right) \tag{1.26}$$

where Δ^{k} is the $k-th$ order forward difference operator.

The Definition of the falling factorial is given by the following definition.

Definition 1.3 The ϑth falling factorial function for any $\vartheta, t \in \mathbb{R}$ is

$$t^{(\vartheta)} = \frac{\Gamma(t+1)}{\Gamma(t+1-\vartheta)}, \quad t \in \mathbb{R}\setminus\{\ldots,-2,-1,0\}. \tag{1.27}$$

if $t+1-\vartheta \in \{0,-1,-2,-3,\cdots\}$ then $t^{(\vartheta)}=0$.

Theorem 1.1 *Let a real valued function χ be defined on N_i and $\lambda, \beta > 0$, then the following*

(i). $\Delta^{-\beta}\left[\Delta^{-\lambda}\chi(\ell)\right] = \Delta^{-(\lambda+\beta)}\chi(\ell) = \Delta^{-\lambda}\left[\Delta^{-\beta}\chi(\ell)\right]$.
(ii). $\Delta^{-\beta}\Delta\chi(\ell) = \Delta\Delta^{-\beta}\chi(\ell) - \frac{(\ell-i)^{(\beta-1)}}{\Gamma(\beta)}\chi(i)$

hold.

Lemma 1.1 *Let $\lambda \neq 1$ and assume $\lambda + \beta + 1$ is a positive integer, then*

$$\Delta^{-\beta} \ell^{(\lambda)} = \frac{\Gamma(\lambda+1)}{\Gamma(\lambda+\beta+1)} \ell^{(\lambda+\beta)}, \qquad \ell \in N_0. \tag{1.28}$$

Lemma 1.2 *Assume that the following factorial functions are well defined*

(i) If $0 < \nu < 1$, then $\ell^{(\nu\gamma)} \geq \left(\ell^{(\gamma)}\right)^{\nu}$
(ii) $\ell^{(\eta+\gamma)} = (\ell - \gamma)^{(\eta)} \ell^{(\gamma)}$.

The Results related to Caputo fractional difference operator are as follows:

Definition 1.4 Let $\mu > 0$ and $k - 1 < \mu < k$, where k denotes a positive integer, $k = \lceil \mu \rceil$, $\lceil . \rceil$ denotes the number ceiling. Set $\alpha = k - \mu$. The $\mu - th$ fractional Caputo like difference is defined as

$$\begin{aligned} \Delta_*^{\mu} g(t) =& \Delta^{-\alpha} \left(\Delta^k g(t)\right) \\ =& \frac{1}{\Gamma(\alpha)} \sum_{s=i}^{t-\alpha} (t-s-1)^{(\alpha-1)} \left(\Delta^k g\right)(s), \quad \forall t \in N_{i+\alpha}, \end{aligned} \tag{1.29}$$

where Δ^k is the $k - th$ order forward difference operator.

Lemma 1.3 *For $\mu > 0$, μ is a non-integer, $k = \lceil \mu \rceil$, $\alpha = k - \mu$, it holds*

$$f(t) = \sum_{m=0}^{k-1} \frac{(t-i)^{(m)}}{m!} \Delta^m [f(a)] + \frac{1}{\Gamma(\mu)} \sum_{s=i+\alpha}^{t-\mu} (t-s-1)^{(\mu-1)} \Delta_*^{\mu} [f(s)],$$

where f is defined on $\mathbb{N}_a$ with $a \in \mathbb{Z}^+$. In particular, if $0 < \mu < 1$ and $a = 0$, then the above relation becomes

$$f(t) = f(0) + \frac{1}{\Gamma(\mu)} \sum_{s=1-\mu}^{t-\mu} (t-s-1)^{(\mu-1)} \Delta_*^{\mu} [f(s)] \tag{1.30}$$

where f is defined on $\mathbb{N}_1$.

Definition 1.5 Let $t \in \mathbb{N}_0$, $\mu \in (-1, 1)$ and $\vartheta \in \mathbb{R}^+$. The discrete mittag leffler function is defined by

$$F_{\vartheta}(\mu, t) = \lim_{a \to \infty} \sum_{j=0}^{a} \mu^j \frac{(t + j(\vartheta - 1))^{(jq)}}{\Gamma(1 + jq)} \tag{1.31}$$

Duffing Map:
The advantages of studying discrete fractional calculus models are demonstrated numerically using the Duffing map. The chaotic Duffing map is given by

$$\begin{cases} u_{n+1} = & v_n \\ v_{n+1} = & -\mu\, u_n + \mu_1 v_n - v_n^3, \end{cases} \tag{1.32}$$

where μ, μ_1 are real constants. The corresponding discrete fractional form of the Duffing equation is given by

$$\begin{cases} \Delta^{\vartheta} u_n = & v_n - u_n \\ \Delta^{\vartheta} v_n = & -\mu\, u_n + (\mu_1 - 1)v_n - v_n^3, \end{cases} \tag{1.33}$$

where Δ^{ϑ} is the Caputo type fractional difference operator of order $0 < \vartheta < 1$. The simulations were conducted using fixed parameter values of $\mu = 0.2$ and $\mu_1 = 2.75$, along with initial conditions of $(0.3, 0.4)$. The chaotic behavior of the system described by equation (1.32) is displayed through time series and phase plane diagrams in Figure 1.5. Figure 1.6 showcases the numerical illustration of the discrete fractional Duffing map given by equation (1.33), where the order is set to $\vartheta = 0.9$. The figures clearly demonstrate that the system exhibits complex dynamic behavior, which varies with changes in the fractional order. Additionally, Figure 1.7 presents time series plots of the system described by equation (1.33), highlighting the transition of the system's state from chaotic to periodic and then back to chaotic as the fractional order ϑ gradually increases within the range of $[0.2, 0.8]$.

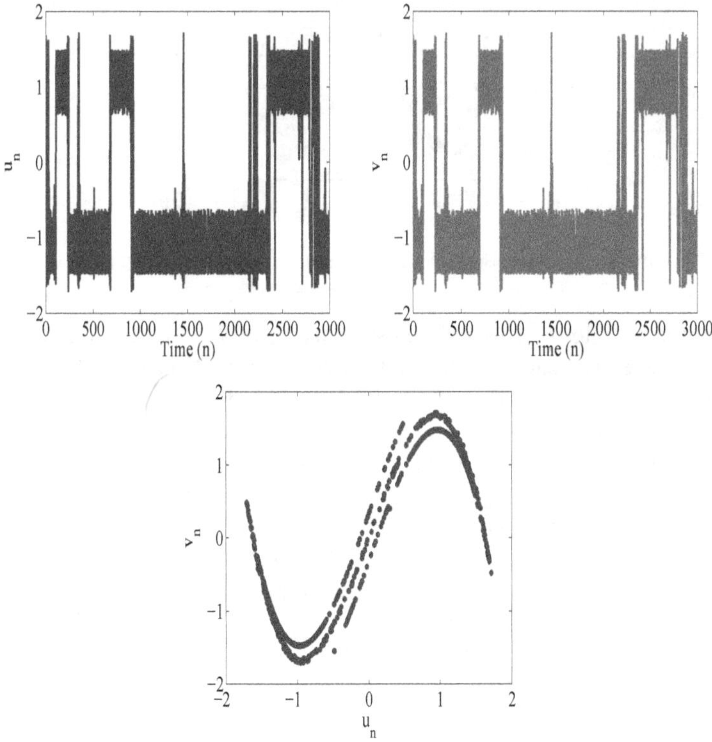

Fig. 1.5: Time series and phase plane for the Duffing map (1.32).

1.4 Chaotic Systems

The term "chaos" is often used to describe a system or event that exhibits no observable variability or pattern in its motion. The concept of chaos has been discussed extensively in the literature. In 1922, Hadamard, while

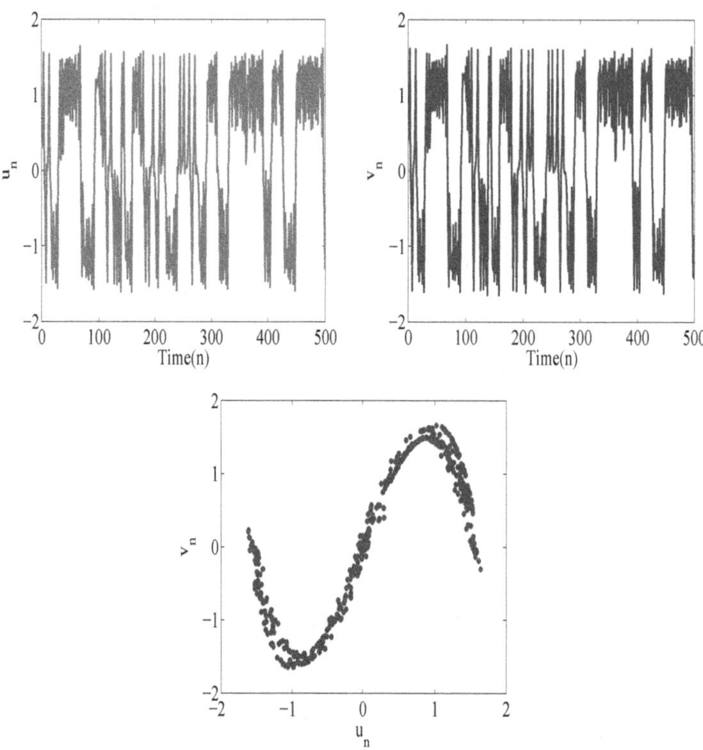

Fig. 1.6: Time series and phase plane for the discrete fractional Duffing map (1.33).

studying partial differential equations, proposed that systems sensitive to initial conditions fail to adhere to the laws of nature. Other scientists such as Maxwell (1876), Henry Poincare (1913), and Duhem (1982) encountered similar phenomena in their respective research. However, it was not until 1963 that Lorenz investigated the emergence of chaos and its practical applications through the concept of the butterfly effect. The butterfly effect refers to the notion that small changes in the environment, such as the flapping of a butterfly's wings, can eventually lead to significant alterations in the path of a tornado or even prevent its occurrence. This concept does not imply a direct cause-and-effect relationship between butterfly wings and tornadoes but rather highlights the interconnectedness of events, where even minor initial changes in the atmosphere can have far-reaching consequences. There continues to be growing interest and curiosity in understanding the impact of chaos in various fields.

Notable applications of chaos include weather forecasting, quantum mechanics, geology, finance, understanding species coexistence, tumor cell growth, robotics, anthropology, and more. Chaos finds application in neural processes, studying heartbeats and blood clot formation, chemical reactions, and encryption of secret messages. Chaotic behavior can be observed in discrete maps, providing insights into the nature of chaos. In 1989, Robert Devaney proposed a powerful definition of chaos for discrete maps, stating that chaos is characterized by a lack of periodicity and weak sensitive dependence of neighboring points in trajectories.

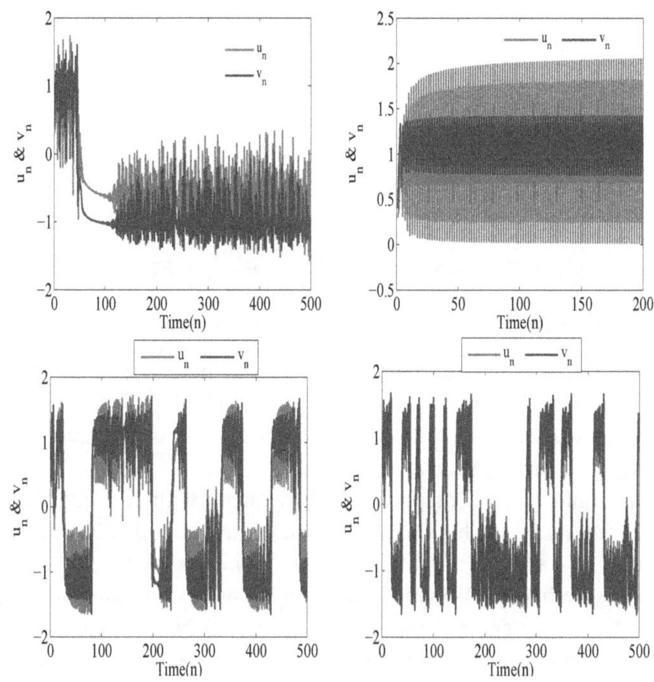

Fig. 1.7: Time series plots for the discrete fractional Duffing map (1.33) for $\vartheta = 0.2, 0.4, 0.6, 0.8$.

1.4.1 Discrete Fractional Chaotic Maps

In the literature, chaotic maps refer to discrete–time dynamical systems that exhibit chaotic behaviour. The most typical discrete maps consist of

1. Logistic map:

$$z_{n+1} = \nu\, z_n(1 - z_n),$$

   where ν is a real parameter.

2. Gauss map:

$$z_{n+1} = e^{-\mu z_n^2} + \mu_1,$$

   where μ, μ_1 are real parameters.

3. Henon map:

$$\begin{cases} z_{n+1} = & 1 - \nu\, z_n^2 + w_n, \\ w_{n+1} = & \nu_1 z_n, \end{cases}$$

   where ν, ν_1 are real constants.

4. Ikeda map: This is a complex map used to model optical resonator given by

$$w_{n+1} = \alpha + \beta w_n e^{iP/(|w_n|^2+1)+\eta},$$

where $\alpha, \beta, P\eta$ are real parameters.

5. Bogdanov map:

$$\begin{cases} w_{n+1} = & w_n + z_{n+1} \\ z_{n+1} = & z_n + \mu\, z_n + \mu_1 w_n(w_n - 1) + \mu_2 w_n z_n, \end{cases}$$

where μ, μ_1, μ_2 are real parameters.

Various maps, such as the sine map, standard map, Baker's map, Horseshoe map, Tangent map, Tinkerbell map, and Chialvo map, have been studied extensively in the context of chaos. Recently, researchers, engineers, and scientists have shown significant interest in discrete fractional versions of these chaotic maps due to their chaotic behavior and wide range of applications.

The field of discrete fractional chaotic maps is relatively new, with the first work published by Wu et al. in 2014, focusing on the study of the sine and standard maps with discrete fractional orders [25, 26]. Subsequently, this idea was extended to investigate the behavior of the logistic map under discrete fractional operators [27]. Discrete fractional maps find applications in communication systems, particularly in encryption algorithms. One advantage of considering fractional order maps is that they introduce an additional parameter in the form of the fractional order. This parameter can be employed in two ways: assuming the same fractional order (commensurate) for all state variables in the system or assuming different fractional orders (incommensurate) for each state variable. This distinction illustrates the physical significance of each state variable and allows for more flexible modeling and analysis.

1.5 Summary

This chapter introduces the application of mathematical modeling in real-world phenomena. It begins with a concise introduction to mathematical models and proceeds to provide illustrative examples of how mathematical models are constructed and applied in various interdisciplinary fields. The chapter also delves into the methods used for constructing these models, aiming to enhance the understanding of the significance of mathematical analysis in real-life situations.The second part of the chapter focuses on the history of fractional calculus and highlights the motivation behind studying fractional order systems. It presents fundamental definitions and theorems in discrete fractional calculus, which will be essential throughout the book. This section aims to establish a solid foundation for the subsequent discussions and analyses related to discrete fractional calculus.

Finally, the chapter addresses chaotic systems and their discrete counterparts. It explores the characteristics and behavior of chaotic systems and their representation in the discrete domain. By considering discrete chaotic systems, the chapter aims to provide insights into the study of chaos from a discrete fractional perspective.

References

[1] Igor Podlubny. Fractional Differential Equations, 198 Academic Press. San Diego, California, USA (1999).
[2] Keith Oldham and Jerome Spanier. The Fractional Calculus Theory and Applications of Differentiation and Integration to Arbitrary Order. Elsevier (1974).
[3] Kenneth S. Miller and Bertram Ross. An Introduction to the Fractional Calculus and Fractional Differential Equations. John Wiley & Sons Inc., New York (1993).
[4] Silvestre François Lacroix. An elementary treatise on the differential and integral calculus. J. Smith (1816).
[5] Nils Henrik Abel. Resolution of a mechanical task. (1826): 153–157.
[6] Joseph Liouville. Mémoire sur quelques questions de géométrie et de mécanique, et sur un nouveau genre de calcul pour résoudre ces questions (1832).
[7] Joseph Liouville. Mémoire sur le théorème des fonctions complémentaires. (1834): 1–19.
[8] Ya. Sonin, N. On differentiation with arbitrary index. Moscow Matem. Sbornik 6, no. 1 (1869): 1–38.
[9] Aleksey Vasilievich Letnikov. Theory of differentiation with an arbtraly indicator. Matem Sbornik 3 (1868): 1–68.
[10] Laurent, H. On the calculus of derivatives with arbitrary indices. New Annals of Mathematics: Journal of Candidates for Polytechnic and Normal Schools 3 (1884): 240–252.
[11] Anton Karl Grunwald. Uber begrente Derivationen und deren Anwedung. Zangew Math und Phys 12 (1867): 441–480.
[12] Oliver Heaviside. Electrical Papers. Vol. 2. Macmillan and Company (1925).
[13] Hermann Weyl. Bemerkungen zum begriff des differentialquotienten gebrochener ordnung. Vierteljschr. Naturforsch. Gesellsch. Zurich 62, no. 1-2 (1917): 296–302.
[14] Godfrey Harold Hardy and John Edensor Littlewood. Some properties of fractional integrals, I. Mathematische Zeitschrift 27, no. 1 (1928): 565–606.
[15] Andre Marchaud. On the derivatives and on the differences of functions of real variables. Journal of Pure and Applied Mathematics 6 (1927): 337–425.
[16] Riesz, M. Intégrale de Riemann-Liouville et solution invariantive du probléme de Cauchy pour l'équation de sondes. In Comptes Rendus du Congres International des Mathematiciens 2 (1936): 44–45.
[17] Michele Caputo. Linear models of dissipation whose Q is almost frequency independent—II. Geophysical Journal International 13, no. 5 (1967): 529–539.
[18] Rudolf Hilfer, ed. Applications of fractional calculus in physics. World Scientific (2000).
[19] Igor Podlubny. Matrix approach to discrete fractional calculus. Fractional Calculus and Applied Analysis 3, no. 4 (2000): 359–386.
[20] Ferhan M. Atici and Paul W. Eloe. A transform method in discrete fractional calculus. International Journal of Difference Equations 2, no. 2 (2007).
[21] George A. Anastassiou. Discrete fractional calculus and inequalities. arXiv preprint arXiv:0911.3370 (2009).
[22] Christopher Goodrich and Allan C. Peterson. Discrete fractional calculus. Vol. 10. Cham: Springer (2015).
[23] Thabet Abdeljawad and Dumitru Baleanu. Fractional differences and integration by parts. Journal of Computational Analysis & Applications 13, no. 3 (2011).

[24] Guo-Cheng Wu and Dumitru Baleanu. Jacobian matrix algorithm for Lyapunov exponents of the discrete fractional maps. Communications in Nonlinear Science and Numerical Simulation 22, no. 1-3 (2015): 95–100.

[25] Guo-Cheng Wu, Dumitru Baleanu and Sheng-Da Zeng. Discrete chaos in fractional sine and standard maps. Physics Letters A 378, no. 5-6 (2014): 484–487.

[26] Guo-Cheng Wu and Dumitru Baleanu. Discrete fractional logistic map and its chaos. Nonlinear Dynamics 75 (2014): 283–287.

[27] Guo-Cheng Wu, Dumitru Baleanu, He-Ping Xie and Fu-Lai Chen. Chaos synchronization of fractional chaotic maps based on the stability condition. Physica A: Statistical Mechanics and its Applications 460 (2016): 374–383.

[28] Qin Fan, Guo-Cheng Wu and Hui Fu. A note on function space and boundedness of the general fractional integral in continuous time random walk. Journal of Nonlinear Mathematical Physics (2022): 1–8.

[29] Hui Fu, Guo–Cheng Wu, Guang Yang and Lan–Lan Huang. Continuous time random walk to a general fractional Fokker–Planck equation on fractal media. The European Physical Journal Special Topics 230, no. 21 (2021): 3927–3933.

[30] Martin Bohner and Allan Peterson. Dynamic Equations on Time Scales: An Introduction with Applications. Springer Science & Business Media (2001).

[31] Kenneth S. Miller and Bertram Ross. Fractional difference calculus. In Proceedings of the International Symposium on Univalent Functions, Fractional Calculus and their Applications (1988): 139–152.

[32] Michael Holm. The theory of discrete fractional calculus: Development and application. The University of Nebraska-Lincoln (2011).

Chapter 2

Chaos and Synchronization of Fractional Order Discrete-Time Chemical Reaction Systems

The process of chemical reactions involves the rearrangement of electrons within molecules, resulting in the formation of new compounds. These reactions are influenced by various environmental factors such as temperature, humidity, and pressure, which can significantly impact their dynamics. This chapter aims to explore and analyze the chaotic behavior and synchronization of chemical reaction systems, as well as their practical applications. Two chemical reaction systems are examined in this chapter using a discrete-time fractional order operator. The first reaction involves the interaction between three components: Malonic acid, chlorine dioxide, and iodine. The second reaction focuses on the application of the ferroelectric effect in neurological waves, based on the Van der Pol oscillator model. Through phase plane analysis, bifurcations, and Lyapunov exponents, the chapter demonstrates the emergence of oscillatory behavior leading to chaos in these two systems. Furthermore, the study extends to the subdivision of subsystems and their synchronization by introducing nonlinear controllers that incorporate feedback gain variables. By exploring the impact of these variables, the investigation reveals intriguing dynamical characteristics of the chemical reaction system and sheds light on the factors that can be controlled to influence the reactions.

2.1 Modelling in Chemistry

Mathematical methods and terminology play a crucial role in understanding both natural and artificial chemical phenomena within the field of chemistry. Mathematics finds application in chemistry in two key ways. Firstly, mathematical methods can be used to calculate the outcomes of chemical reactions, providing quantitative results. Secondly, mathematical models can be constructed to explain and address various chemical problems, enabling the interpretation of facts and the answering of related questions.

Chemical reactions can be represented and studied as mathematical models that consider the reactants, products, and the energy factors between them. By obtaining solutions for these mathematical models, we can gain insights into the behavior of the chemical reaction. Representing chemical reactions mathematically offers advantages such as revealing hidden behaviors that may not be observed during experiments. This allows experimental designs to be developed specifically to observe these predicted behaviors, thereby validating the theoretical models. The theoretical construction of models for chemical reactions is cost-effective and time-saving. However, the success of this approach depends on accurately modeling the reaction kinetics. Mathematical methods generally exhibit high accuracy rates for modeling small molecules, while modeling large molecules requires significant computational time and system memory.

One intriguing aspect of chemical reactions is their oscillatory behavior, which has several applications. Initially, there were limited studies on oscillatory chemical reactions due to experimental design constraints, and computational approaches were primarily focused on reproducing observed experimental results. However, advancements in technology and a curiosity to explore new dimensions of study have led to investigations of theoretical models under slight parameter variations and their impact on the dynamics of constructed chemical reaction models. While most chemical reactions are unidirectional under specific conditions, there are rare cases where reactions exhibit reversibility and continue to oscillate repeatedly. It's important to note that not all chemical reactions display oscillatory behavior. When oscillating, reactions can exhibit different phenomena such as sequential color changes, foam formation, gas bursts, and more.

Yatsimirskii et al. (1982) introduced a novel type of chemical oscillations involving copper and nickel, and their findings are presented in their publication [7]. Epstein et al. conducted a study on the oscillations of chemical reactions, and their work can be found in [5]. The dynamics of these oscillations were further investigated by Field et al. in 1983 [6]. Taylor (2002) provides a comprehensive discussion on the important mechanisms and processes involved in oscillating chemical reactions [8]. Argoul (1987) reported the initial study of chaos in chemical reactions [10]. Field (1993) extensively illustrates the chaotic behavior exhibited by oscillating reaction models in his book [9].

2.1.1 Oscillatory Reactions: A Physical Interpretation

When discussing oscillations, the common analogy provided is that of a pendulum. However, the oscillatory behavior of chemical reactions is fundamentally different from the motion of a pendulum. This distinction arises due to the contradiction between the oscillatory reactions and the second law of thermodynamics. According to the second law of thermodynamics, a chemical reaction that reaches equilibrium cannot spontaneously shift from that position. In the case of oscillatory reactions, they occur far from equilibrium, and the decrease in energy of the compounds formed drives the reaction to reverse its direction. This mechanism, involving the reverse reaction due to a decrease in compound energy, is one of the three attributes of oscillatory reactions. Another attribute is the formation of two pathways, which are periodically switched due to energy released during the reactions. The final attribute involves the formation of a new compound that acts as a stimulus, facilitating the transition between the two pathways.

The development of the field of chemical dynamics is largely attributed to the capability of mathematical models to replicate experimental observations. While replicating experimental results is not an easy task, numerical simulations can be constructed by adjusting various parameters to achieve desired outcomes. Differential equation models are widely accepted as a feasible means of representing reaction systems, and qualitative analysis is conducted by understanding the solutions of these constructed differential equations. These models not only aid in analytically studying and reproducing experimental results but also provide an avenue for exploring unknown domains and further enhancing scientific studies. One significant advantage of modeling is its ability to consider a large number of parameters that influence or play a vital role in chemical reactions, including complex or multi-step reactions. Field (2015) illustrates chaos in the BZ reactor model, further highlighting the intricacies of oscillating reactions [11].

One significant property of chemical reactions is their ability to exhibit "hysteresis" or a "memory effect." This refers to the change in the flow of chemical reactions when the flow rate reaches a certain value, causing the chemical reaction system to retain a memory of the flow. While the concept of memory is more commonly used among physicists in modeling, it is relatively rare among chemists. However, certain processes such as adsorption and autocatalysis demonstrate the memory effect.

Recent advancements in mathematical modeling, particularly involving arbitrary order derivatives, provide a better description of the hysteresis property compared to classical differential equation systems. The introduction of fractional derivatives has emerged as a more accurate and illustrative approach for capturing the dynamic nature of constructed chemical reaction systems. Several recent studies have focused on fractional order chemical reaction models. For example, Owolabi et al. [16] constructed a BZ model with an Atangana-type fractional derivative. Sarwar et al. [17] conducted stability analysis of nonlinear chemical reaction systems with arbitrary order derivatives, exploring their dynamical properties. The effect of diffusion in fractional chemical systems was studied by Jena et al. in 2020 [18], while Veersha (2022) implemented a numerical approach to study reaction mechanisms [19].

In the realm of chaotic dynamics in fractional order systems, only a few authors have contributed thus far. Vijay et al. [20] conducted a qualitative study with the introduction of a control term in fractional order systems, while He et al. investigated coexisting attractors in such systems [21]. Notably, there have been no studies on chemical reaction models using discrete fractional order operators. To address this research gap and explore the complex dynamics, this chapter focuses on chemical systems utilizing discrete–time fractional order operators.

2.1.2 Oscillating Chemical Reactions

The initial discovery of oscillating chemical reactions can be traced back to Fechner's experiment, where an oscillating current was produced by electrochemical cells. However, the existence of these oscillating reactions remained largely theoretical until the early 20th century. It was in the 1970s that the concept of nonlinearity in chemical reactions gained wide acceptance and established itself. Several well-known models of oscillating reactions have been developed, including Lotka's model, the Oregonator, and the Brusselator. Lotka's model, originally proposed to explain chemical oscillations, was later adapted by Volterra to discuss ecological phenomena involving interactions between different species populations. Autocatalysis is a prominent feature of both Lotka-Volterra models and chemical reaction systems that exhibit oscillations. Autocatalysis refers to a growth rate that depends on the increasing concentration of a species, leading to oscillatory behavior among the species' concentrations.

The study of oscillatory chemical reactions has been prevalent among chemists since 1945. Describing the mechanism of a chemical reaction involves step-by-step illustrations of the reaction pathway leading to the overall outcome. However, providing a mechanism is not always an easy task, especially for complex chemical reactions. In the following section, we will discuss some interesting and important examples of oscillating chemical reactions.

Belousov-Zhabotinsky Reaction (BZ reaction):

The study of the Belousov-Zhabotinsky (BZ) reaction has attracted the attention of physicists, chemists, and mathematicians due to its diverse range of applications. For instance, the BZ reaction serves as a prototype for understanding chemical reactions in the cardiac system, showcasing excitability and wave-like behavior in both systems. The overall BZ reaction is

$$3CH_2(COOH)_2 + 4BrO_3^- \longrightarrow 4Br^- + 9CO_2 + 6H_2O. \tag{2.1}$$

Cerium is the most commonly used catalyst in the BZ reaction, but other catalysts such as cobalt, chromium, manganese, osmium, nickel, and silver are also employed in this context. The mechanism of the BZ reaction is highly complex, involving a large number of pathways. Over time, new pathways have been included based on experimental evidence. However, it is worth noting that many of the initial findings on chemical oscillatory reactions were published in the Russian language, which posed a challenge for scientists in the Western world who did not have access to those publications. Furthermore, during the Cold War and due to various political issues at the time, limited exposure and communication between Eastern and Western scientists hindered the dissemination of knowledge on these reactions.

The Oscillating Clock:
The Briggs-Rauscher reaction, also known as the oscillating clock reaction, is renowned for its striking color changes. The reaction begins by mixing three colorless solutions, which initially turns to an amber color. Suddenly, the color shifts to a dark blue hue, which then gradually fades away. This process repeats, ultimately culminating in a deep purple color. The dynamics of the system can be understood by observing the step-by-step changes in color, sometimes with instantaneous shifts. The formation of the amber color is attributed to the production of free iodine, and when the reaction ceases, an immediate change to a purple color can be observed. The overall reaction is

$$IO_3^- + 2H_2O_2 + CH_2(COOH)_2 + H^+ \rightarrow ICH(COOH)_2 + 2O_2 + 3H_2O. \tag{2.2}$$

2.2 Chemical Reaction Systems

The discovery and experimental confirmation of oscillating chemical reactions occurred only in the late 1970s, thanks to the efforts of chemical engineers worldwide. These reactions were often stumbled upon accidentally, as no common methods were initially proposed for intentionally generating chemical oscillators. During this time, only two oscillatory reactions, namely the Belousov-Zhabotinsky (BZ) reaction and the Bray reaction, were widely accepted. As researchers began to discover new chemical oscillators by manipulating the catalysts used in reactions, interest in analyzing and understanding the behavior of these oscillatory systems grew rapidly. Subsequently, attention also turned towards investigating oscillatory behavior in biological systems. In this chapter, we delve into the analysis of both chemical oscillator models and biological oscillator models, particularly those related to neurology. The focus lies on studying the chaotic dynamics of these systems and exploring synchronization using nonlinear oscillators. Before delving into the mathematical analysis of these systems, we provide a brief overview of the Lengyel-Epstein chemical reaction and the biological snap oscillator, which exhibits a ferroelectric effect.

2.2.1 Lengyel-Epstein Model of Chemical Reactions

In this section, we explore the Lengyel-Epstein model, which was put forth in 1991 to describe the chemical reactions involving iodine, chlorine dioxide, and malonic acid. These reactions serve as a direct representation of the model. Interestingly, this model's findings provide confirmation of the pattern formation study conducted by Alan Turing in 1952. The chemical reaction scheme is given by

$$\begin{aligned} MA + I_2 &\rightarrow I\,MA + I^- + H^+, \\ ClO_2 + I^- &\rightarrow \frac{1}{2} I_2 + ClO_2^-, \\ ClO_2^- + 4I^- + 4H^+ &\rightarrow Cl^- + 2I_2 + 2H_2O. \end{aligned} \tag{2.3}$$

The first step in the scheme (2.3) involves the iodization of malonic acid (MA). The second step is the oxidation of iodine ions with radicals of ClO_2, while the final step represents the formation of iodine from chlorite and iodine ions produced in the previous two reactions. In reactions (2.3), the concentrations of iodine and chlorite ions fluctuate, while the concentrations of MA and ClO_2 change very slowly. These slow changes can be treated as constant factors, while the fluctuations in iodine and chlorine ion concentrations can be regarded as variables for constructing a differential equation model of the aforementioned scheme. By employing the rate law, the given chemical reactions in (2.3) can be converted into a dimensionless mathematical form as follows:

$$\begin{cases} x'(t) = & \zeta_1 - x(t) - \dfrac{4x(t)y(t)}{1+x(t)^2}, \\ y'(t) = & \zeta_2 x(t)\left[1 - \dfrac{y(t)}{1+x(t)^2}\right]. \end{cases} \tag{2.4}$$

The concentration of chlorite ions (ClO_2^-) and iodine ions (I^-) are represented by $x(t)$ and $y(t)$, respectively. The constants ζ_1 and ζ_2 correspond to the values of the feed concentrations.

To investigate the oscillatory reaction between chlorine dioxide, malonic acid, and iodine, the mathematical system is analyzed after discretizing using the forward Euler method and using the fractional Caputo difference operator, which is given by the following expressions.

$$\begin{cases} \Delta^{\vartheta_1} x(n) = & \zeta_1 - x(n-1+\vartheta_1) - \dfrac{4x(n-1+\vartheta_1)y(n-1+\vartheta_1)}{1+(x(n-1+\vartheta_1))^2}, \\ \Delta^{\vartheta_2} y(n) = & \zeta_2 x(n-1+\vartheta_2)\left(1 - \dfrac{y(n-1+\vartheta_2)}{1+(x(n-1+\vartheta_2))^2}\right), \end{cases} \tag{2.5}$$

where Δ^{ϑ_1} is the Caputo difference operator with $0 < \vartheta_1, \vartheta_2 \leq 1$, $\zeta_1, \zeta_2 > 0$ are real parameters and $n \in \mathbb{N}_{1-\vartheta_i}, i = 1, 2$. Employing the method of fractional sum equation [2] to (2.5) for conversion to a numerical form given by

$$\begin{cases} x(n) = & x(0) + \dfrac{1}{\Gamma(\vartheta_1)} \sum\limits_{k=1}^{n} \dfrac{\Gamma(n-k+\vartheta_1)}{\Gamma(n-k+1)} \left(\zeta_1 - x(j-1) - \dfrac{4x(j-1)y(j-1)}{1+(x(j-1))^2}\right), \\ y(n) = & y(0) + \dfrac{1}{\Gamma(\vartheta_2)} \sum\limits_{k=1}^{n} \dfrac{\Gamma(n-k+\vartheta_2)}{\Gamma(n-k+1)} \left(\zeta_2 x(j-1)\left(1 - \dfrac{4x(j-1)y(j-1)}{1+(x(j-1))^2}\right)\right), \quad n = 1, 2, 3\cdots. \end{cases} \tag{2.6}$$

2.2.2 Enzyme and Substrate Mechanism Model

Oscillatory chemical reactions play a crucial role in various organisms, starting from cyanobacteria to mammals like humans. The study of chemical reaction systems would be incomplete without discussing biochemical reaction systems, as they are essential for maintaining the physiological functions of biological systems. Cellular oscillations are involved in processes such as glucose formation, adenosine monophosphate generation, protein and DNA signaling, and more. Additionally, physiological oscillations encompass vital processes like heartbeats and respiration. In biological systems, the occurrence of chaos signifies drastic changes that can potentially lead to the death of an individual. Therefore, understanding the chaotic behavior of biological systems and finding effective ways to control chaos becomes a significant area of research for biologists and mathematicians.

The Van der Pol oscillator model is commonly used to represent various physiological oscillations in humans. It has been applied to describe reactions among substrates and enzymes in the brain and the rhythmic beating of the heart. In this chapter, we present an autonomous mathematical model of substrate and enzyme interaction exhibiting ferroelectric behavior without external stimulation. The model is represented by a first-order system, given by,

$$\begin{cases} \dot{x} = & y, \\ \dot{y} = & \delta\, y\left(1 - x^2 + \delta_1 x^4 - \delta_2 x^6\right) - x, \end{cases} \tag{2.7}$$

where, $\delta, \delta_1, \delta_2$ are real constants. The system (2.7) is discretized using Caputo difference operator (Δ^{ϑ_i}) where $0 < \vartheta_i < 1$ and $n \in \mathbb{N}_{1-\vartheta_i}, i = 1, 2$ is given by,

$$\begin{cases} \Delta^{\vartheta_1} x(n) = & y(n-1+\vartheta_1), \\ \Delta^{\vartheta_2} y(n) = & \delta\, y(n-1+\vartheta_2)\left(1 - x(n-1+\vartheta_2)^2 + \delta_1 x(n-1+\vartheta_2)^4 - \delta_2 x(n-1+\vartheta_2)^6\right) \\ & -x(n-1+\vartheta_2), n = 0, 1, 2, \cdots. \end{cases} \tag{2.8}$$

The numerical form of system (2.8) is,

$$\begin{cases} x(n) = & x(0) + \dfrac{1}{\Gamma(\vartheta_1)} \sum\limits_{k=1}^{n} \dfrac{\Gamma(n-k+\vartheta_1)}{\Gamma(n-k+1)} \left(y(j-1)-\right), \\ y(n) = & y(0) + \dfrac{1}{\Gamma(\vartheta_2)} \sum\limits_{k=1}^{n} \dfrac{\Gamma(n-k+\vartheta_2)}{\Gamma(n-k+1)} \Big(\delta\, y(j-1)\left(1 - x(j-1)^2 + \delta_1 x(j-1)^4 - \delta_2 x(j-1)^6\right) \\ & -x(j-1)\Big), \; n = 1, 2, 3, \cdots. \end{cases} \tag{2.9}$$

For detailed information on mathematical analysis of fractional order systems with discrete time and Lyapunov exponents for discrete fractional maps readers can refer to [1, 2, 3, 4].

2.3 Chaotic Dynamics of Lengyel-Epstein Model

Experimental investigations are often conducted to understand the mechanisms of chemical reactions and make theoretical predictions. However, it is not always possible to fully prove a mechanism, as sometimes it may fail to align with the predicted outcomes. In such cases, the mechanism needs to be reconstructed or modified to restart the process. This is where the dynamic analysis of an equivalent mathematical model plays a crucial role, as it provides insights into the long-term behavior of the chemical mechanism being studied.

In this section, we explore the chaotic behavior exhibited by the Lengyel-Epstein Model using bifurcation diagrams and Lyapunov exponents. The discussion is divided into two subsections. The first subsection focuses on the system with the same fractional order, while the second subsection considers different orders for the system of equations. These analyses will shed light on the chaotic dynamics of the Lengyel-Epstein Model and contribute to a better understanding of its behavior.

2.3.1 Commensurate Order System

The system (2.5) is referred to as a commensurate order system when $\vartheta_1 = \vartheta_2$. Commensurate order systems are commonly used to analyze the dynamics of the system. By fixing the values of other system parameters, the influence of a parameter variation on the system dynamics can be investigated. The analysis of system bifurcation aids in qualitatively understanding the behavior of the system.

Numerical Validation: The bifurcation diagram and largest Lyapunov exponent of the system (2.5) with the same fractional order and for the following values of the parameters $\vartheta_1 = 0.9$, $\zeta_2 = 1.2$ and varying $\zeta_1 \in [6, 16]$ with initial condition $(0.01, 0.02)$ are visualized in Figure 2.1.

The time series plots corresponding to the bifurcation diagram in Figure 2.1 provide insights into the changes in the system states for different values of ζ_1. The occurrence of chaos in the system described by (2.5) is evident from both the bifurcation diagram and the positive values of the Lyapunov exponents shown in

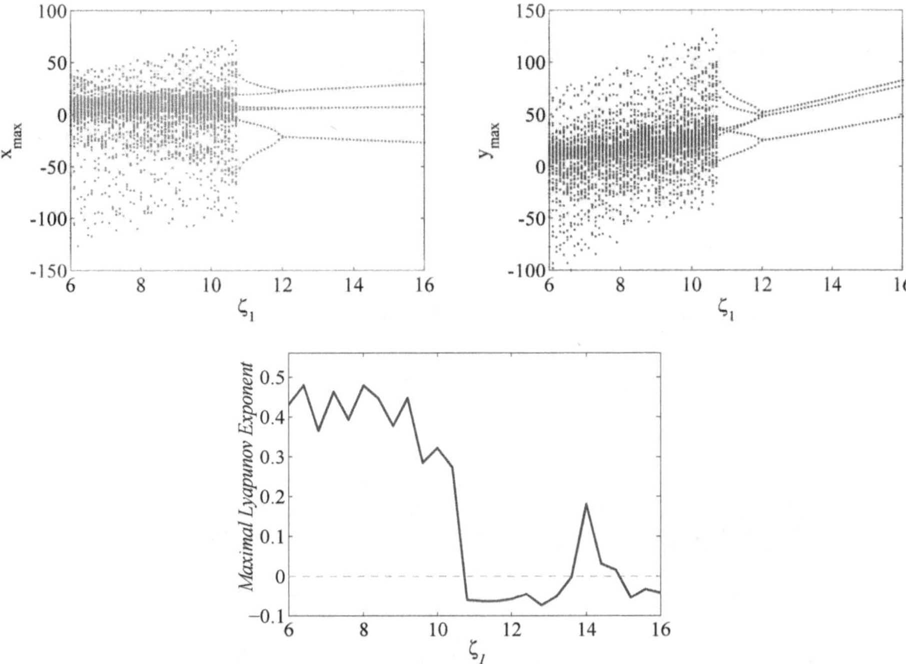

Fig. 2.1: Bifurcation diagram and Lyapunov exponent of system (2.5) with $\vartheta_1 = \vartheta_2$.

Figure 2.1. By observing Figure 2.1, it can be noted that the system exhibits aperiodic oscillations for values of ζ_1 ranging from 6 to 11. After this range, the time series trajectories demonstrate uniform oscillations. Hence, for the given parameter values, the system's behavior becomes increasingly oscillatory over time.

To illustrate the changes in the system's behavior, time series plots are provided for the state variables at various values of ζ_1. Figure 2.2 displays trajectories for $\zeta_1 = 6$ and 8, Figure 2.3 shows trajectories for $\zeta_1 = 10$ and 12, while Figure 2.4 depicts trajectories for $\zeta_1 = 14$ and 16. These plots highlight the unsteady nature of the model and further emphasize the dynamic behavior observed in the system.

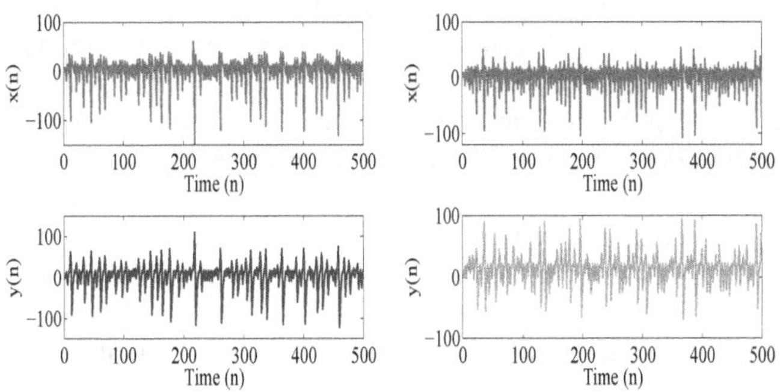

Fig. 2.2: Time series plots corresponding to bifurcation diagram and Lyapunov exponent of system $\vartheta_1 = \vartheta_2$ (2.5) in Figure 2.1.

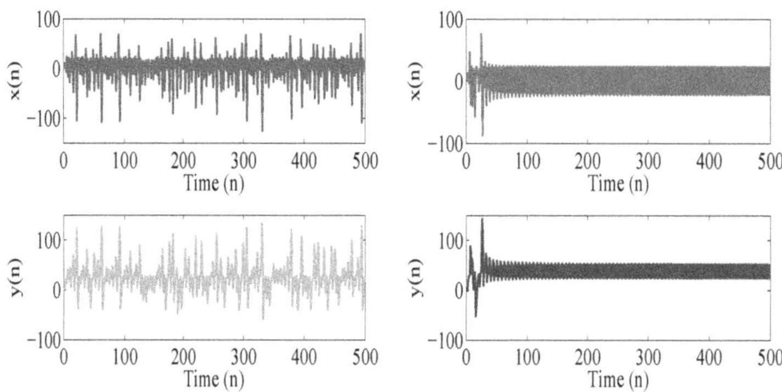

Fig. 2.3: Time series plots corresponding to bifurcation diagram and Lyapunov exponent of system (2.5) with $\vartheta_1 = \vartheta_2$ in Figure 2.1.

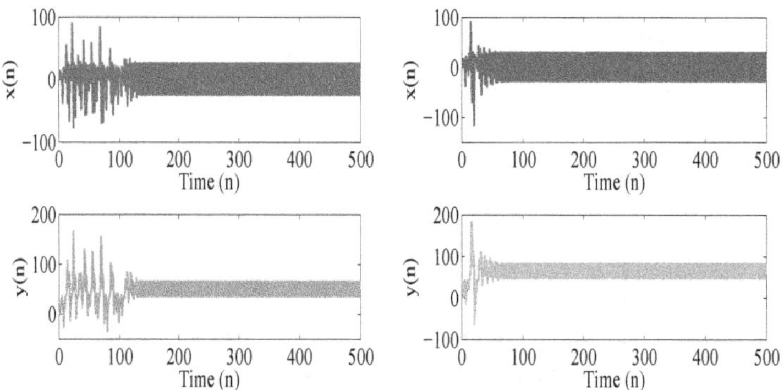

Fig. 2.4: Time series plots corresponding to bifurcation diagram and Lyapunov exponent of system (2.5) with $\vartheta_1 = \vartheta_2$ in Figure 2.1.

2.3.2 Incommensurate Fractional Order System

This section of the chapter explores the consequences of considering different orders for the state variables that describe a chemical reaction model. The purpose of studying incommensurate orders is to uncover hidden characteristics of the systems. In order to compare the behavior of the system under commensurate and incommensurate orders, numerical simulations are conducted using the same parameter values as in the previous section, with the only difference being the use of different fractional orders ($\vartheta_1 \neq \vartheta_2$). To demonstrate the chaotic behavior of the system described by Equation (2.5) under different fractional orders, bifurcation diagrams and corresponding Lyapunov exponents are presented in Figure 2.5. The parameter values chosen for these simulations are $\zeta_2 = 1.2$, $\vartheta_1 = 0.95$, $\vartheta_2 = 0.45$, and the parameter ζ_1 is in the range of 6 to 16. The initial condition used for the simulations is $(0.01, 0.02)$.

When considering different fractional orders, the system continues to exhibit chaotic behavior similar to the commensurate order case, as shown in Figure 2.5. However, there are noticeable variations in the system's behavior, which can be observed from the corresponding Lyapunov exponents. In the case of incommensurate order, the maximum Lyapunov exponent is found to be 0.2846, whereas it is 0.4915 for the commensurate

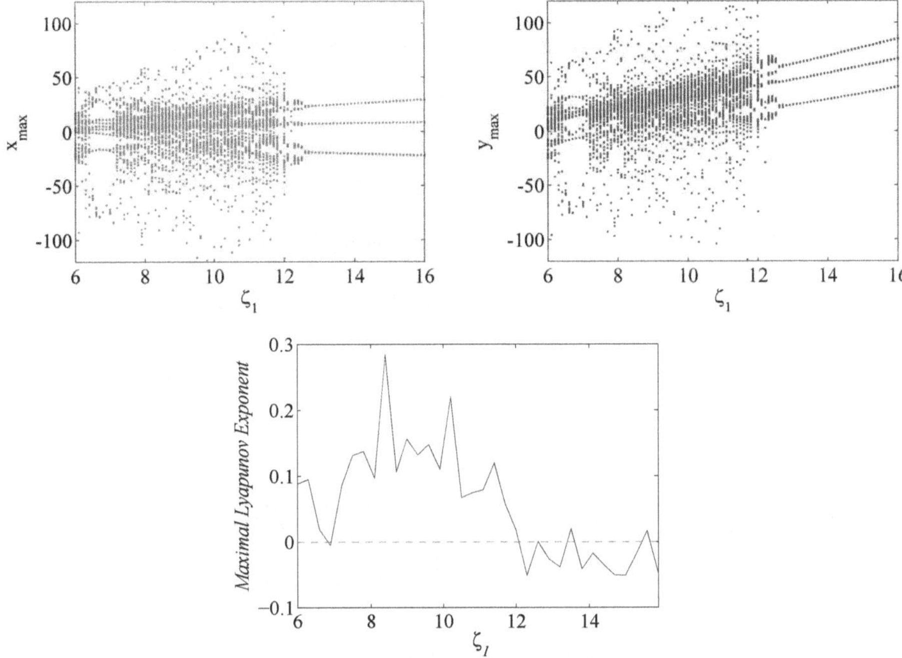

Fig. 2.5: Bifurcation diagram and Lyapunov exponent of system (2.5).

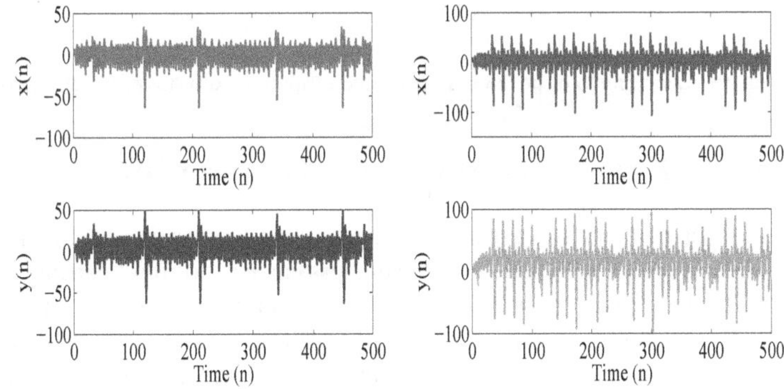

Fig. 2.6: Time series plots corresponding to bifurcation diagram and Lyapunov exponent of system (2.5) in Figure 2.5.

order case. Additionally, for higher values of ζ_1, the fluctuation in the system's behavior is minimized under incommensurate order conditions.

This analysis highlights the significance of the fractional order of the state variables in understanding the behavior of the system. In this subsection, the fractional order of the state variable $y(n)$ is assumed to be $\vartheta_2 = 0.45$, which is comparatively smaller than the fractional order of the state variable $x(n)$, assumed to be $\vartheta_1 = 0.95$. Consequently, the decrease in the order of the second state variable, which represents the concentration of chlorite ions, has a significant impact on the system's behavior and acts as a factor in controlling the occurrence of chaos.

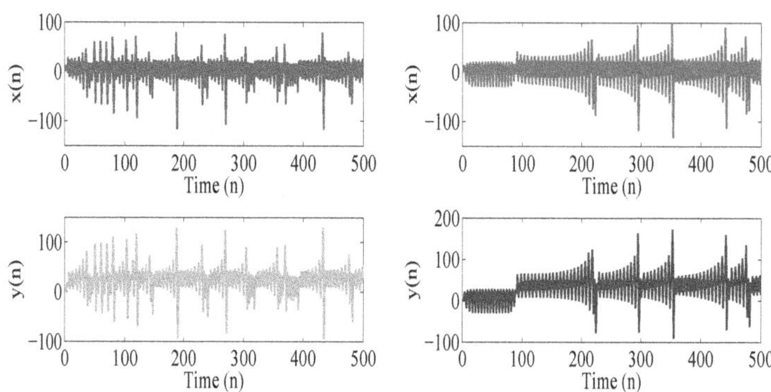

Fig. 2.7: Time series plots corresponding to bifurcation diagram and Lyapunov exponent of system (2.5) in Figure 2.5.

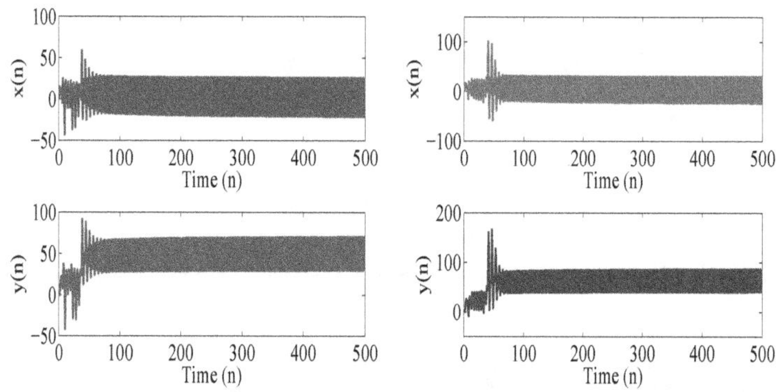

Fig. 2.8: Time series plots corresponding to bifurcation diagram and Lyapunov exponent of system (2.5) in Figure 2.5.

2.4 Chaotic Dynamics of Enzyme and Substrate Mechanism Model

The analysis of an enzyme-substrate model is essential for comprehending various mechanisms in biological systems. Even minor changes in the factors involved in the process can lead to significant and often intolerable consequences, sometimes resulting in death. In this section, we investigate a biological snap oscillator model and explore the chaotic behavior exhibited by the system. Specifically, we focus on examining the effect of the damping term coefficient in the Van der Pol oscillator, which is used for modeling purposes. By conducting a study, we aim to demonstrate the impact of this coefficient on the system's behavior.

To illustrate these effects, we employ bifurcation diagrams, which depict the changes in the system's dynamics, and examine the transition of the state variables from stable to unstable states. Furthermore, we employ Lyapunov exponents to quantify the degree of instability and chaos observed in the system. By analyzing these measures, we can gain insights into how variations in the damping term coefficient influence the overall behavior of the oscillator model.

2.4.1 Commensurate Order System

The dynamical behavior of the system is demonstrated with bifurcation diagrams, and the unpredictable behavior of the system (2.8) is presented with Lyapunov exponents. The values of the parameters are as follows: $\delta_1 = 0.1$, $\delta_2 = 0.22$ and the order of the system $\vartheta_1 = \vartheta_2 = 0.25$ with the initial condition $(0.1, 0.2)$. The simulations carried out with varying $\delta \in (0, 0.8)$ are visualized in Figure 2.9.

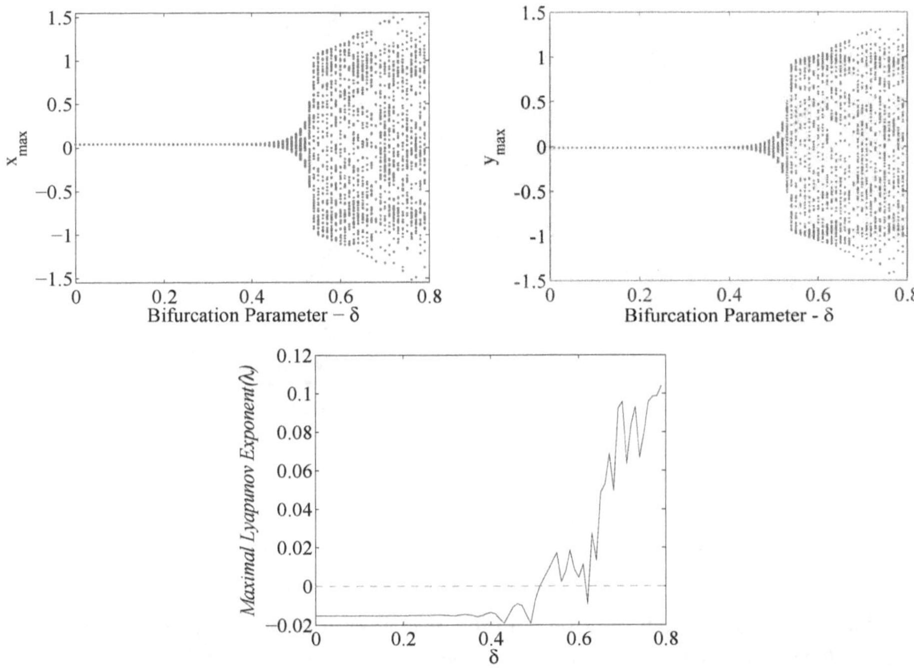

Fig. 2.9: Bifurcation diagram and Lyapunov exponent of system (2.5).

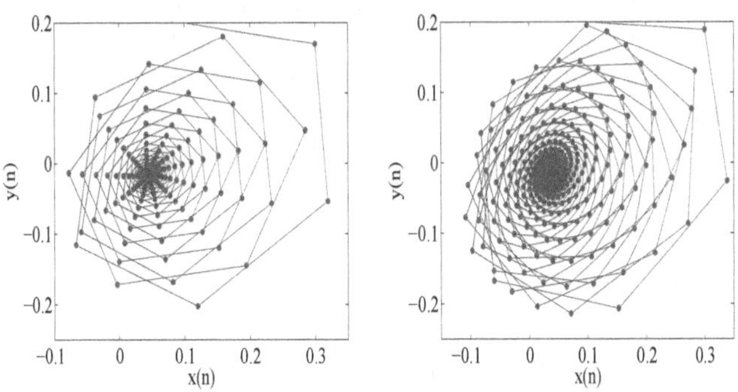

Fig. 2.10: Phase portrait corresponding to bifurcation diagram and Lyapunov exponent of system (2.8) in Figure 2.9.

The initial values of δ result in a straight line, indicating that the system exhibits stable behavior. Over time, the system tends to approach its equilibrium position, indicating a steady-state behavior. This behavior is visually portrayed in the phase plane plot shown in Figure 2.10, where trajectories starting from initial

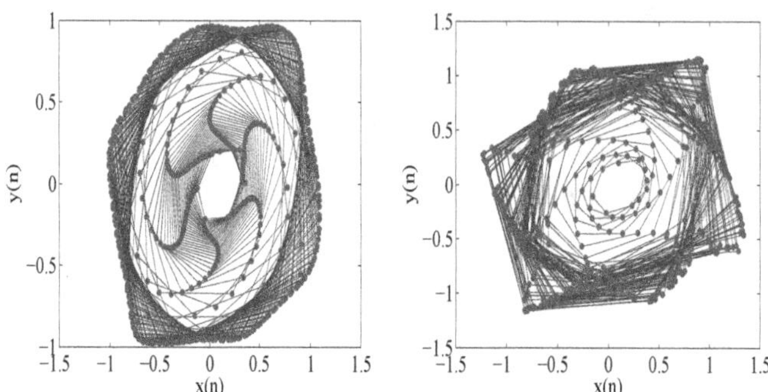

Fig. 2.11: Phase portrait corresponding to bifurcation diagram and Lyapunov exponent of system (2.8) in Figure 2.9.

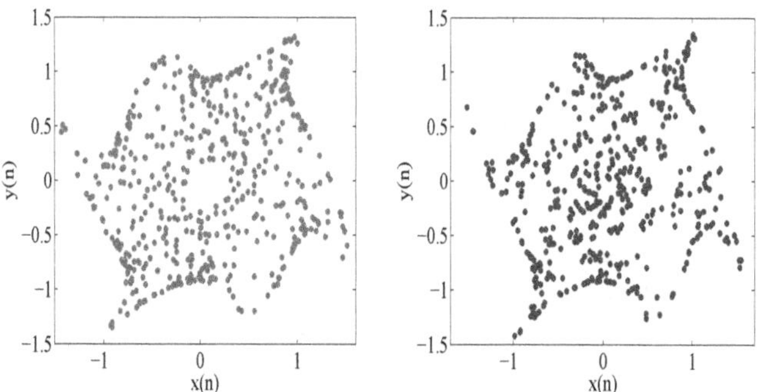

Fig. 2.12: Phase portrait corresponding to bifurcation diagram and Lyapunov exponent of system (2.8) in Figure 2.9.

positions move towards the equilibrium position with the formation of a spiral. The transition from a stable state to a limit cycle formation state is demonstrated in Figure 2.11 for $\delta = 0.55$ and 0.65. In these cases, the phase plane plot shows trajectories that no longer converge to a single point but instead form closed loops. Figure 2.12 illustrates the aperiodic trajectory motion or random oscillatory behavior of the system. The trajectory does not exhibit a repetitive pattern and appears to move erratically.

The behavior of the system, including the stable, oscillatory, and chaotic nature, is further characterized by the Lyapunov exponents shown in Figure 2.9. Negative values of the Lyapunov exponents correspond to stable and uniform oscillatory behavior, while positive values indicate a chaotic nature in the system. These Lyapunov exponents provide quantitative insights into the dynamics and stability of the system under different parameter values.

2.4.2 Incommensurate Order System

For different fractional orders of the system (2.8) $\vartheta_1 = 0.2$ and $\vartheta_2 = 0.1$ with $\delta_1 = 0.1$, $\delta_2 = 0.22$, initial condition $(0.1, 0.2)$, the bifurcation diagram and Lyapunov exponents with varying $\delta \in (0, 0.86)$ are presented in Figure 2.13. Similar to the study of the chemical reaction model in the previous section, the occurrence of chaos is compared between the state variables with the same fractional orders and those with different fractional orders.

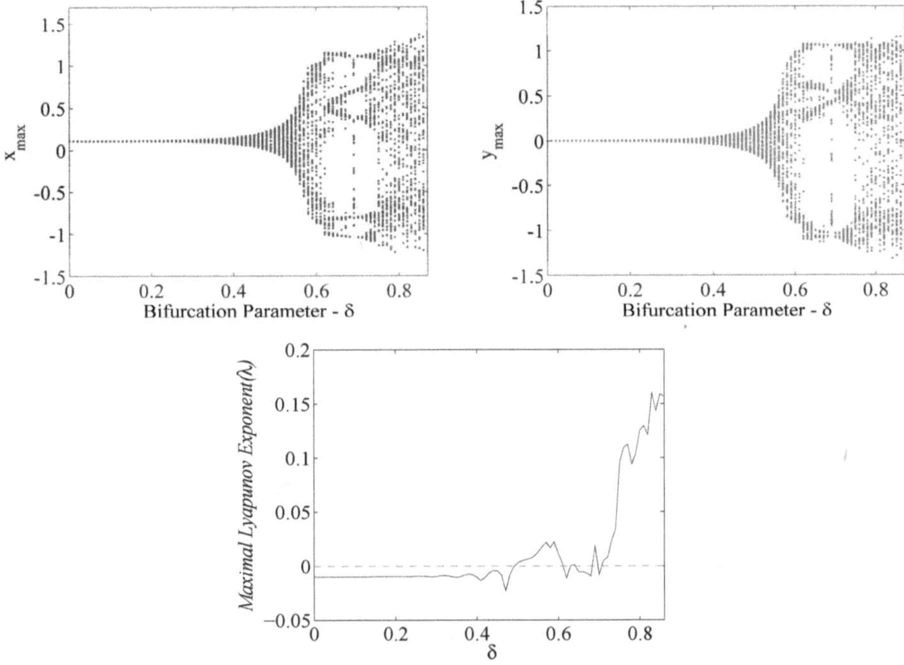

Fig. 2.13: Bifurcation diagram and Lyapunov exponent of system (2.5).

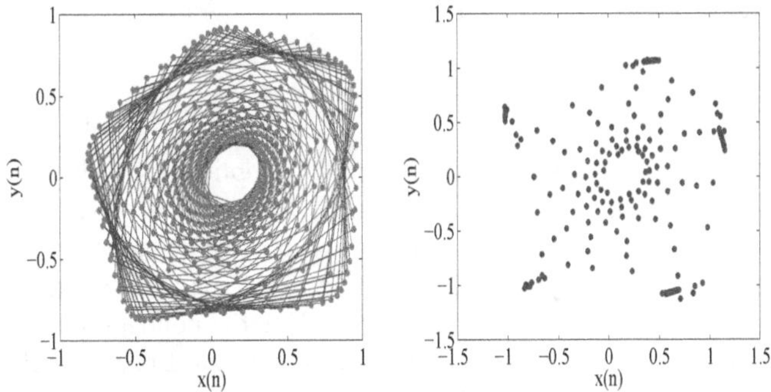

Fig. 2.14: Phase portrait corresponding to bifurcation diagram and Lyapunov exponent of system (2.8) in Figure 2.9.

When considering the system described by Equation (ER), the bifurcation diagram exhibits a noticeable shift when the fractional orders ϑ_1 and ϑ_2 are at their minimum values. This shift indicates a change in the system's dynamics and behavior. Moreover, the chaotic behavior of the system extends up to a value of δ equal to 0.86, resulting in a prolonged period of chaos. This implies that the system remains in a chaotic state for a longer duration, exhibiting complex and unpredictable behavior. By incorporating different fractional orders, unexpected behaviors of the system are brought to light. These variations in the orders of the system's variables unveil novel dynamics and characteristics, showcasing the intricate nature of the system's behavior.

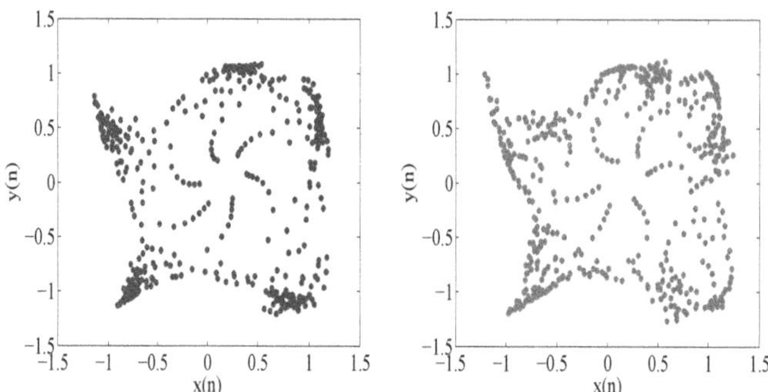

Fig. 2.15: Phase portrait corresponding to bifurcation diagram and Lyapunov exponent of system (2.8) in Figure 2.9.

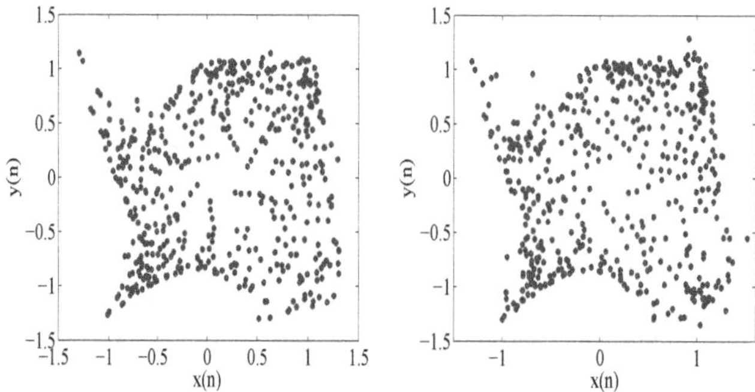

Fig. 2.16: Phase portrait corresponding to bifurcation diagram and Lyapunov exponent of system (2.8) in Figure 2.9.

2.5 Synchronization

The high sensitivity of chaotic systems is crucial for many real-life applications, which has led to an increased interest in studying the synchronization of these chaotic systems. Various synchronization schemes have been introduced, including adaptive control, feedback control, active and nonlinear control, among others. While most synchronization schemes are typically applied to identical systems or subsystems, there is also active progress in research to synchronize different chaotic systems. Notable contributions in the field of control methods and synchronization of chemical systems include an article by Bodale on the synchronization of the Willamowski-Rössler system [12], an experimental analysis by Guderian et al. studying electrochemical-based chaos control [13], and the synchronization of the perturbed Briggs-Rauscher system by Kumpinsky et al. [22]. It was in the 21st century that synchronization of chemical systems began to gain attention and evolve. Optimal type synchronization was introduced by Shabunin et al. in 2003 [14], and waveform illustration and synchronization were discussed by Tanaka et al. [15].

Fractional order synchronization has garnered attention due to its ability to provide better descriptions of real-life phenomena. The literature on synchronization in fractional order models is vast, with a large number of contributions. Zhang et al. proposed three schemes for the synchronization of fractional order chaotic systems in [23], while Wu et al. introduced the idea of synchronization to discrete fractional systems in [24]. You et al. presented a neural network model constructed with fractional order and discrete–time for synchronization in [25].

In this section, nonlinear controllers are employed to synchronize the considered chemical reaction systems described by Equations (1) and (ER). By utilizing these controllers, the goal is to achieve synchronization between the two systems, enabling them to exhibit coordinated behavior.

2.5.1 Synchronization of Lengyel-Epstein Model

The synchronization of the Lengyel-Epstein model involves the formation of master and slave subsystems, which are then coupled together using nonlinear controllers with feedback gain. The aim is to achieve synchronization between the two subsystems, indicating that their behaviors become coordinated and exhibit similar dynamics.

For synchronization to be achieved, the error dynamical system, which represents the difference between the master and slave subsystems, should be asymptotically stable. This means that the error between the two subsystems diminishes over time, eventually reaching a state of synchronization. The stability of the error dynamical system is assessed for the specific parameter values considered in the study.
The master and slave systems are given by,

$$\begin{cases} \Delta^{\vartheta_1} x_M(n) = & \zeta_1 - x_M(n-1+\vartheta_1) - \dfrac{4x_M(n-1+\vartheta_1)y_M(n-1+\vartheta_1)}{1+(x_M(n-1+\vartheta_1))^2}, \\ \Delta^{\vartheta_2} y_M(n) = & \zeta_2 x_M(n-1+\vartheta_2)\left(1 - \dfrac{4x_M(n-1+\vartheta_2)y_M(n-1+\vartheta_2)}{1+(x_M(n-1+\vartheta_2))^2}\right), \end{cases} \tag{2.10}$$

and

$$\begin{cases} \Delta^{\vartheta_1} x_S(n) = & \zeta_1 - x_S(n-1+\vartheta_1) - \dfrac{4x_S(n-1+\vartheta_1)y_S(n-1+\vartheta_1)}{1+(x_S(n-1+\vartheta_1))^2} + W_1(n-1+\vartheta_1), \\ \Delta^{\vartheta_2} y_S(n) = & \zeta_2 x_S(n-1+\vartheta_2)\left(1 - \dfrac{y_S(n-1+\vartheta_2)}{1+(x_S(n-1+\vartheta_2))^2}\right) + W_2(n-1+\vartheta_2), \end{cases} \tag{2.11}$$

where $(x_M, y_M)^T$ and $(x_S, y_S)^T$ are the master and slave states, $n \in \mathbb{N}_{1-\vartheta_i}, i = 1,2$. The nonlinear control functions are denoted by $W_1(t)$ and $W_2(t)$. The states of error between master and slave systems are defined by

$$(e_1, e_2)^T = (x_S, y_S)^T - (x_M, y_M)^T. \tag{2.12}$$

The corresponding error dynamical system takes the following form

$$\begin{cases} \Delta^{\vartheta_1} e_1(n) = & -e_1(n-1+\vartheta_1) - 4\left(\dfrac{x_S(n-1+\vartheta_1)y_S(n-1+\vartheta_1)}{1+(x_S(n-1+\vartheta_1))^2} - \dfrac{x_M(n-1+\vartheta_1)y_M(n-1+\vartheta_1)}{1+(x_M(n-1+\vartheta_1))^2}\right) \\ & +W_1(n-1+\vartheta_1), \\ \Delta^{\vartheta_2} e_2(n) = & \zeta_2 e_1(n-1+\vartheta_2) - \zeta_2\left(\dfrac{x_S(n-1+\vartheta_2)y_S(n-1+\vartheta_2)}{1+(x_S(n-1+\vartheta_2))^2} - \dfrac{x_M(n-1+\vartheta_2)y_M(n-1+\vartheta_2)}{1+(x_M(n-1+\vartheta_2))^2}\right) \\ & +W_2(n-1+\vartheta_2), \end{cases} \tag{2.13}$$

Nonlinear control functions be defined by

$$\begin{aligned} W_1(t) &= 4\Theta - r_1 e_1 \\ W_2(t) &= \zeta_2\Theta - r_2 e_2 \end{aligned} \tag{2.14}$$

where $\Theta = \left(\frac{x_M x_S\,[e_2 x_s] + e_1 e_2 + x_S y_M + x_m(y_S - 2y_M)}{(1+x_S)^2(1+x_M)^2}\right)$, $0 < r_1, r_2 < 1$ are the control gains. The reduced error system (2.5) is

$$\begin{cases} \Delta^{\vartheta_1} e_1(n) = & -(r_1+1)e_1(n-1+\vartheta_1), \\ \Delta^{\vartheta_2} e_2(n) = & \zeta_2 e_1(n-1+\vartheta_2) - r_2 e_2(n-1+\vartheta_2). \end{cases} \tag{2.15}$$

The time series evolution of the error of synchronization are presented in two cases.

Case-I:

The system constitutes the same fractional order for both state variables. For $\vartheta_1 = \vartheta_2 = 0.99$, $r_1 = 0.9$, $r_2 = 0.9$, $\zeta_2 = 1.2$ with initial condition $(0.1, 0.2)$, the time evolution error plot is presented in Figure 2.18. The time series evolution of the state variables before the introduction of the control terms is presented for the commensurate order case in Figure 2.17. The convergence of the error state variables to the origin as $n \longrightarrow \infty$ can be visualized from Figure 2.18.

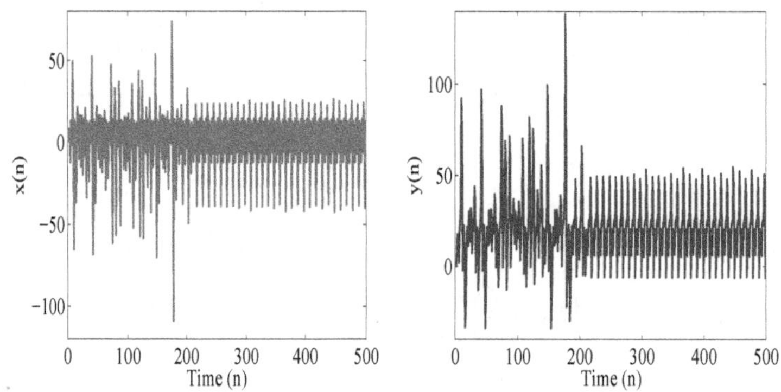

Fig. 2.17: Time evolution of uncontrolled commensurate order form of system (2.5).

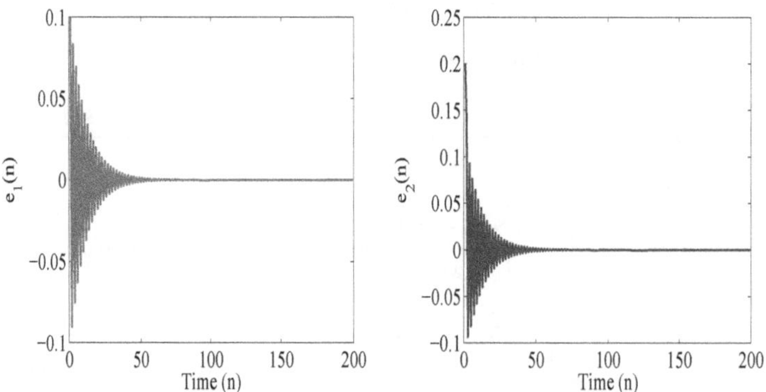

Fig. 2.18: Time evolution of synchronization error.

Case-II:

In this case, the system with state variables taking up different fractional orders is considered. For $\vartheta_1 = 0.99$, $\vartheta_2 = 0.45$, $r_1 = 0.95$, $r_2 = 0.95$, $\zeta_2 = 1.2$ with the initial condition $(0.1, 0.2)$, the time evolution error plot is presented in Figure 2.20. A time series plot of state variables of the uncontrolled system (2.5) with distinct fractional order is demonstrated with Figure 2.19.

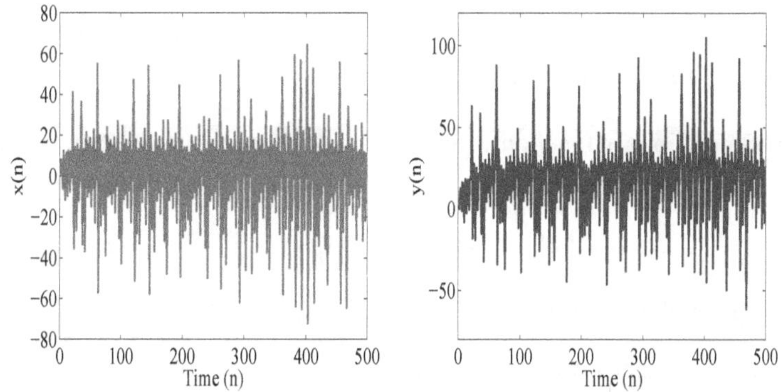

Fig. 2.19: Time evolution of uncontrolled commensurate order form of system (2.5).

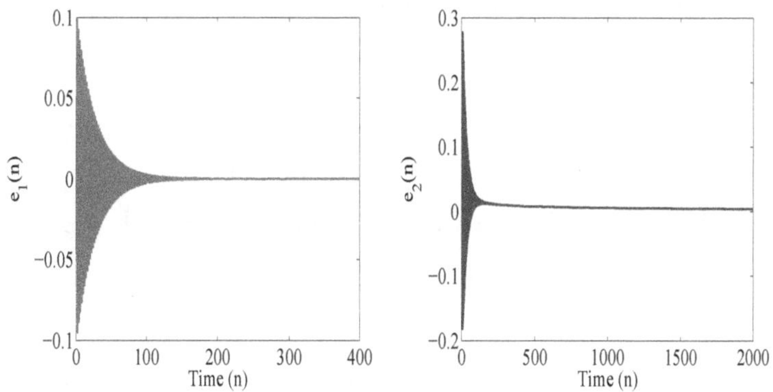

Fig. 2.20: Time evolution of synchronization error.

2.5.2 Synchronization of Enzyme–Substrate Mechanism Model

In this subsection of the chapter, we explore the synchronization of biological snap oscillators, which has potential applications in neurological wave phenomena. Specifically, we focus on synchronizing the enzyme and substrate mechanism model described by Equation (ER) using nonlinear controllers. To achieve synchronization, we divide the system into two subsystems: a master system and a slave system. The master system serves as the reference system, while the slave system is the target system to be synchronized. By implementing appropriate nonlinear controllers, we aim to establish a coordinated behavior between the master and slave systems. This synchronization approach provides insights into the dynamics of biological snap oscillators and opens up possibilities for understanding and controlling neurological wave phenomena, which play important roles in various biological processes.

$$\begin{cases} \Delta^{\vartheta_1} x_D(n) = & y_D(n-1+\vartheta_1), \\ \Delta^{\vartheta_2} y_D(n) = & \delta\, y_D(n-1+\vartheta_2)\left(1 - x_D(n-1+\vartheta_2)^2 + \delta_1 x_D(n-1+\vartheta_2)^4 - \delta_2 x_D(n-1+\vartheta_2)^6\right) \\ & -x_D(n-1+\vartheta_2), \end{cases} \tag{2.16}$$

and

$$\begin{cases} \Delta^{\vartheta_1} x_R(n) = & y_R(n-1+\vartheta_1) + \Phi_1(n-1+\vartheta_1), \\ \Delta^{\vartheta_2} y_R(n) = & \delta\, y_R(n-1+\vartheta_2)\left(1 - x_R(n-1+\vartheta_2)^2 + \delta_1 x_R(n-1+\vartheta_2)^4 - \delta_2 x_R(n-1+\vartheta_2)^6\right) \\ & -x_R(n-1+\vartheta_2) + \Phi_2(n-1+\vartheta_2), \end{cases} \tag{2.17}$$

where Φ_1, Φ_2 are the nonlinear control functions, where master and slave states are represented by $(x_D, y_D)^T$ and $(x_R, y_R)^T$, $n \in \mathbb{N}_{1-\vartheta_i}, i = 1, 2$. Then the error states will be defined as,

$$(e_x, e_y)^T = (x_R, y_R)^T - (x_D, y_D)^T. \tag{2.18}$$

The error dynamical system is obtained as

$$\begin{cases} \Delta^{\vartheta_1} e_x(n) = & e_y(n-1+\vartheta_1) + \Phi_1(n-1+\vartheta_1), \\ \Delta^{\vartheta_2} e_y(n) = & \delta\, e_y(n-1+\vartheta_2) - e_x(n-1+\vartheta_2) + \delta y_R(n-1+\vartheta_2)\big(-x_R(n-1+\vartheta_2)^2 \\ & +\delta_1 x_R(n-1+\vartheta_2)^4 - \delta_2 x_R(n-1+\vartheta_2)^6\big) + \Phi_2(n-1+\vartheta_2) \\ & -\delta\, y_D(n-1+\vartheta_2)\big(-x_D(n-1+\vartheta_2)^2 + \delta_1 x_D(n-1+\vartheta_2)^4 \\ & -\delta_2 x_D(n-1+\vartheta_2)^6\big). \end{cases} \tag{2.19}$$

Let the nonlinear functions be defined by

$$\begin{aligned} \Phi_1(n) &= -\eta_1 e_x(n) \\ \Phi_2(n) &= \delta\, y_D(n)\left(-x_D(n)^2 + \delta_1 x_D(n)^4 - \delta_2 x_D(n)^6\right) \\ &\quad - \delta y_R(n)\big(-x_R(n)^2 + \delta_1 x_R(n)^4 \\ &\quad - \delta_2 x_R(n)^6\big) - \eta_2 e_y(n), \end{aligned} \tag{2.20}$$

with feedback gain functions η_1 and η_2. After implementing the control functions, the reduced system is

$$\begin{cases} \Delta^{\vartheta_1} e_x(n) = & e_y(n-1+\vartheta_1) - \eta_1 e_x(n-1+\vartheta_1), \\ \Delta^{\vartheta_2} e_y(n) = & (\delta - \eta_2)\, e_y(n-1+\vartheta_2) - e_x(n-1+\vartheta_2). \end{cases} \tag{2.21}$$

The time series of synchronization errors are demonstrated in two cases

Case-I:
In this particular case, we analyze the synchronization of the system described by Equation (2.8) using the same fractional order values, $\vartheta_1 = \vartheta_2 = 0.27$. Figure 2.21 illustrates the aperiodic behavior of the system for specific parameter values, namely $\delta = 0.7$, $\delta_1 = 0.1$, and $\delta_2 = 0.22$. The plot demonstrates the erratic and unpredictable nature of the system's behavior under these conditions. On the other hand, when considering feedback control, the system achieves synchronization for the given values of feedback gains, $\eta_1 = 1$ and

$\eta_2 = 1$. Figure 2.22 shows the converging time series plot, where the system's trajectories approach the origin, indicating the successful synchronization of the master and slave systems.

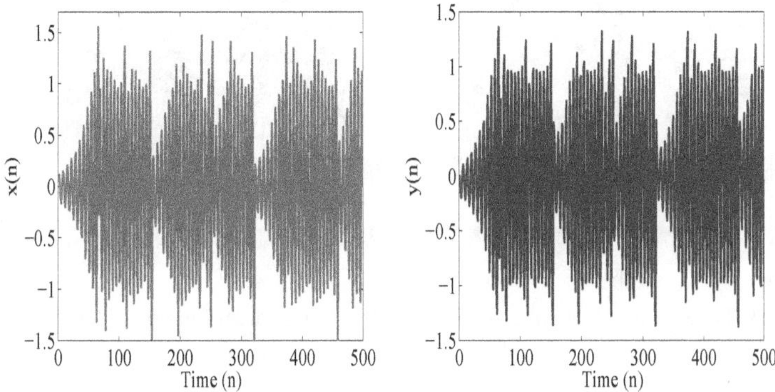

Fig. 2.21: Time evolution of uncontrolled commensurate order form of system (2.8).

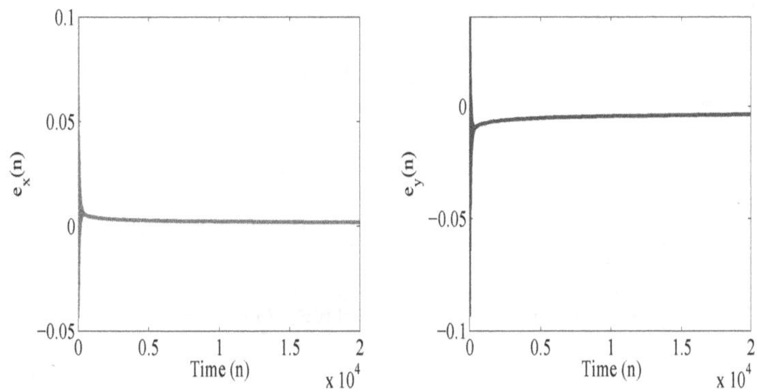

Fig. 2.22: Time evolution of synchronization error.

Case-II:

In this case, we consider the system described by Equation (ER) with different fractional orders: $\vartheta_1 = 0.3$ and $\vartheta_2 = 0.25$. Figure 2.23 depicts the behavior of the system without any control functions, using parameter values of $\delta = 0.7$, $\delta_1 = 0.1$, and $\delta_2 = 0.22$. The time series evolution plot showcases the system's dynamics under these conditions, illustrating its complex and aperiodic behavior. To achieve synchronization, we utilize feedback control with specific feedback gains. Figure 2.24 shows the converging time series plot of the error state variables, as described by Equation (2.21), for feedback gains $\eta_1 = 1.02$ and $\eta_2 = 1$. The plot demonstrates the convergence of the error variables, indicating successful synchronization between the master and slave systems. By incorporating appropriate feedback control mechanisms and adjusting the feedback gains, we can effectively synchronize the system with different fractional orders, leading to coordinated dynamics between the subsystems.

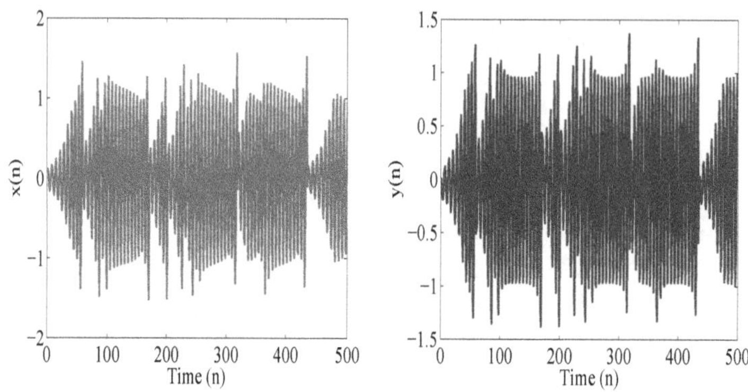

Fig. 2.23: Time evolution of uncontrolled incommensurate order form of system (2.8).

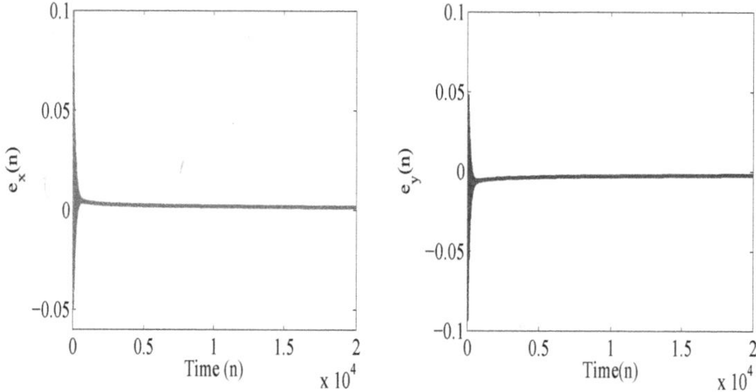

Fig. 2.24: Time evolution of synchronization error.

2.6 Summary

This chapter provides a comprehensive overview of the hysteresis effects observed in chemical reaction systems. It focuses on constructing discrete-time models of chemical reactions with fractional orders and investigates their chaotic nature. Two specific chemical reaction systems have been studied: a three-step reaction system and a biochemical system related to neurological waves, specifically using the snap oscillator model without external effects. The chapter presents time series plots and phase portraits to illustrate the transitions of the system states between periodic and aperiodic states. The influence of a single parameter on the overall system behavior is analyzed through bifurcation diagrams, and regions of chaos are identified using Lyapunov exponents. Furthermore, the chapter compares the behavior of commensurate and incommensurate order systems under the same set of parameter values but with varying fractional orders. This comparison highlights the impact of fractional orders on the system dynamics. The final section of the chapter focuses on the synchronization of subsystems, represented as master and slave versions, using nonlinear controllers with feedback gain functions. Numerical results are presented for the synchronized systems both with and without the inclusion of control terms. Overall, this chapter emphasizes the importance of constructing and investigating chemical kinetics using fractional order operators in discrete-time models, showcasing their relevance in understanding hysteresis effects and system behaviors.

References

[1] Jan Cermak, Istvan Gyori and Ludek Nechvatal. On explicit stability conditions for a linear fractional difference system. Fractional Calculus and Applied Analysis 18 (2015): 651–672.

[2] Adel Ouannas, Amina Aicha Khennaoui, Shaher Momani, Giuseppe Grassi and Viet-Thanh Pham. Chaos and control of a three-dimensional fractional order discrete-time system with no equilibrium and its synchronization. AIP Advances 10, no. 4 (2020): 045310.

[3] Mohd Taib Shatnawi, Noureddine Djenina, Adel Ouannas, Iqbal M. Batiha and Giuseppe Grassi. Novel convenient conditions for the stability of nonlinear incommensurate fractional-order difference systems. Alexandria Engineering Journal 61, no. 2 (2022): 1655–1663.

[4] Guo-Cheng Wu and Dumitru Baleanu. Jacobian matrix algorithm for Lyapunov exponents of the discrete fractional maps. Communications in Nonlinear Science and Numerical Simulation 22, no. 1-3 (2015): 95–100.

[5] Irving R. Epstein, Kenneth Kustin, Patrick De Kepper and Miklos Orban. Oscillating chemical reactions. Scientific American 248, no. 3 (1983): 112–123.

[6] Richard J. Field and F. W. Schneider. Oscillating chemical reactions and nonlinear dynamics. Journal of Chemical Education 66, no. 3 (1989): 195.

[7] Yatsimirskii, K. B. L. P. Tikhonova, L. N. Zakrevskaya, Ya D. Lampeka and A. G. Kolchinskii. New oscillating chemical reactions involving copper and nickel tetraaza macrocyclic complexes. Reaction Kinetics and Catalysis Letters 21, no. 3 (1982): 381–386.

[8] Annette F. Taylor. Mechanism and phenomenology of an oscillating chemical reaction. Progress in Reaction Kinetics and Mechanism 27, no. 4 (2002): 247–326.

[9] Richard J. Field. Chaos in chemistry and biochemistry. World Scientific (1993).

[10] Francoise Argoul, A. Arneodo, P. Richetti, J. C. Roux and Harry L. Swinney. Chemical chaos: From hints to confirmation. Accounts of Chemical Research 20, no. 12 (1987): 436–442.

[11] Richard J. Field. Chaos in the Belousov–Zhabotinsky reaction. Modern Physics Letters B, 29, no. 34 (2015): 1530015.

[12] Ilie Bodale and Victor Andrei Oancea. Chaos control for Willamowski–Rossler model of chemical reactions. Chaos, Solitons & Fractals 78 (2015): 1–9.

[13] Guderian, A., A. F. Munster, M. Kraus and F. W. Schneider. Electrochemical chaos control in a chemical reaction: experiment and simulation. The Journal of Physical Chemistry A 102, no. 26 (1998): 5059–5064.

[14] Shabunin, A., Vladimir Astakhov, Valentin Demidov, Astero Provata, Florence Baras, Gregoire Nicolis and Vadim Anishchenko. Modeling chemical reactions by forced limit-cycle oscillator: Synchronization phenomena and transition to chaos. Chaos, Solitons & Fractals 15, no. 2 (2003): 395–405.

[15] Hisa-Aki Tanaka, Isao Nishikawa, Jurgen Kurths, Yifei Chen and Istvan Z. Kiss. Optimal synchronization of oscillatory chemical reactions with complex pulse, square, and smooth waveforms signals maximizes Tsallis entropy. EPL (Europhysics Letters) 111, no. 5 (2015): 50007.

[16] Kolade M. Owolabi and Zakia Hammouch. Spatiotemporal patterns in the Belousov–Zhabotinskii reaction systems with Atangana–Baleanu fractional order derivative. Physica A: Statistical Mechanics and its Applications 523 (2019): 1072–1090.

[17] Sarwar, S. and S. Iqbal. Stability analysis, dynamical behavior and analytical solutions of nonlinear fractional differential system arising in chemical reaction. Chinese Journal of Physics 56, no. 1 (2018): 374–384.

[18] Rajarama Mohan Jena, Snehashish Chakraverty, Hadi Rezazadeh and Davood Domiri Ganji. On the solution of time-fractional dynamical model of Brusselator reaction-diffusion system arising in chemical reactions. Mathematical Methods in the Applied Sciences 43, no. 7 (2020): 3903–3913.

[19] Veeresha, P. The efficient fractional order based approach to analyze chemical reaction associated with pattern formation. Chaos, Solitons & Fractals 165 (2022): 112862.

[20] Vijay K. Yadav, Subir Das, Beer Singh Bhadauria, Ashok K. Singh and Mayank Srivastava. Stability analysis, chaos control of a fractional order chaotic chemical reactor system and its function projective synchronization with parametric uncertainties. Chinese Journal of Physics 55, no. 3 (2017): 594–605.

[21] Shaobo He, Santo Banerjee and Kehui Sun. Complex dynamics and multiple coexisting attractors in a fractional-order microscopic chemical system. The European Physical Journal Special Topics 228, no. 1 (2019): 195–207.

[22] Enio Kumpinsky, Irving R. Epstein and Patrick De Kepper. Model study of synchronization and other phenomena in light perturbation of the Briggs–Rauscher reaction. International Journal of Chemical Kinetics 17, no. 3 (1985): 345–354.

[23] Yanbin Zhang and Tianshou Zhou. Three schemes to synchronize chaotic fractional-order Rucklidge systems. International Journal of Modern Physics B, 21, no. 12 (2007): 2033–2044.

[24] Guo-Cheng Wu, Dumitru Baleanu, He-Ping Xie and Fu-Lai Chen. Chaos synchronization of fractional chaotic maps based on the stability condition. Physica A: Statistical Mechanics and its Applications 460 (2016): 374–383.

[25] Xingxing You, Qiankun Song and Zhenjiang Zhao. Global Mittag-Leffler stability and synchronization of discrete-time fractional-order complex-valued neural networks with time delay. Neural Networks 122 (2020): 382–394.

Chapter 3

Chaos in Discrete-Time Fractional Order Bio-Chemical Models

The construction of mathematical models for natural phenomena and events is a highly complex task, and scientists and researchers in the field of science and technology are actively working to gain a comprehensive understanding of the workings of nature. While many processes are intricate, some can be represented analytically. Biochemistry, as an interdisciplinary field, focuses on understanding the functioning of organisms at the cellular and molecular levels. In this chapter, the primary goal is to develop discrete-time fractional-order models of biochemical reactions that have significant applications in human life. Two distinct chemical reactions of biological importance are chosen for the analysis of chaos. Firstly, a nonlinear mathematical model of the two-step reversible reactions involved in glucose isomerization in the presence of borate ions is considered. This reaction finds wide-ranging applications in biofuels and the generation of hydrocarbons. Secondly, the focus is on a reaction closely related to the biological mechanism of the cell cycle in eukaryotic organisms. The cell cycle system is developed by considering the active forms of molecules responsible for the occurrence of the cell cycle. Initially, mathematical models are constructed using differential equations of integer order. Subsequently, these models are converted into discrete-time fractional-order systems to analyze their chaotic behavior. The chaotic response of the systems is demonstrated through bifurcation diagrams accompanied by Lyapunov exponents. To support the breakdown analysis of the bifurcation diagrams, a series of time series plots and phase plane analyses are conducted. The study examines two scenarios: one involving a system with identical fractional orders for all state variables, and the other involving non-identical fractional orders of the state variables. By analyzing these bio-chemical systems, researchers gain insights into the complex phenomena of nature and the functioning of biological organisms, particularly with regard to bio-switches.

3.1 Introduction

Chemical reactions play a vital role in the functioning and survival of living organisms. These reactions, also known as metabolism, encompass various processes such as photosynthesis in plants, digestion, energy conversion from food, waste excretion, and transportation between cells. They are essential for growth, organ function, reproduction, and other biological activities. These reactions can be categorized as catabolic (breaking down compounds) and anabolic (synthesizing compounds). In any biological species, reactions occur in a series of pathways, forming a complete metabolic process. The product of one reaction becomes the reactant for the next reaction, and so on.

Several factors, including temperature, affect these chemical processes. Additionally, reactants don't always spontaneously collide to initiate a reaction, necessitating the presence of catalysts. Enzymes, acting as catalysts, not only facilitate rapid bio-chemical reactions but also play a significant role in regulating cellular responses to changes in the environment. There are various types of bio-chemical reactions, such as isomerization, hydrolysis, ligation, transfer, redox, and carbon bond formation/removal.

For instance, hydrogenation involves adding hydrogen atoms to carbon bonds, commonly observed in the production of fatty acids from vegetable oils. Ligation reactions join molecules together using energy derived from ATP. Hydrolysis processes involve the removal of water as a byproduct. Redox reactions are crucial for providing energy to cells through the conversion of food substances with inhaled oxygen during respiration. Thus, bio-chemical processes form the foundation for the proper functioning and survival of living organisms. Among these reactions, redox reactions are particularly critical as they serve as the major energy source for cellular metabolism.

Understanding oscillations in chemical and biochemical reactions is highly beneficial for constructing practical applications in real-life situations [4], [5]. Any changes in the surrounding environment have an immediate impact on biological systems, prompting them to respond accordingly. These responses occur through the transfer of signals across the systems. The study of biological systems in the 20th century was primarily experimental, with a focus on genetics and molecular biology. However, in the early 21st century, the increasing influence of physics and mathematics revolutionized the field of biology. Advancements in technology have also played a significant role in facilitating interdisciplinary studies and the evaluation of data using computational algorithms.

3.2 Mathematical Models of Bio-Chemical Reactions

Maintaining the steady state of biochemical reactions is crucial for the survival of living organisms. Even small disturbances in these reactions can lead to the collapse of cellular metabolism and potentially result in fatality. However, biological systems have mechanisms in place to self-regulate and respond to external disturbances by either accelerating or decelerating reactions as needed. Understanding the behavior of these reactions is essential, and mathematical models are ideal tools for their investigation. The advancement of biotechnology emphasizes the importance of comprehending reaction dynamics, as it opens doors to developments in the fields of biotechnology and medicine, enabling the treatment of diseases and exploration of unknown behaviors.

Various research contributions have focused on the modeling and chaos analysis of biochemical reactions. Numerical methods have been employed for the analysis of biochemical models in studies such as [20], [21], and [22], while topological analysis has been performed in [27]. Bifurcation analysis of biochemical reactions with nonlinearity has been conducted in [15], [26], [28], and [29]. The existence of Bogdanov-Takens bifurcation in a biological chemical process model was established in [16]. Bifurcations with impulsive perturbations were explored in [23], and practical implications were proposed in [25]. The discrete-time version of chemical models and flip bifurcation were investigated in [19]. Additionally, mathematical models incorporating fractional derivatives were studied in [17], [18], and [24] for biochemical reactions.

Motivated by the need for analytical studies of biochemical processes and the investigation of their chaotic nature using fractional order discrete-time operators, this chapter aims to develop mathematical models. Specifically, it focuses on the isomerization of sugar molecules in the presence of borate ions and the cell cycle model in fission yeast.

3.2.1 Model Formulation of Borate Ions Binding with Sugar

The production of renewable fuel resources and biofuels has become a subject of increasing interest. The understanding of biological reactions plays a crucial role in developing this ideology. Applications of biochemical reactions in renewable energy resources can be found in [13], while the formation of glucose in the presence of borate is illustrated in [14]. Glucose and xylose are essential components of carbohydrates. Researchers have explored the generation of hydrocarbons by processing a mixture of hexose and pentose. In this context, our particular focus is on investigating the chaotic response of glucose isomerization in the presence of borate ions. This chemical reaction involves two steps, with the binding of tetrahydroxyborate ions to two sugar molecules [12]. The presence of borate ions enhances the isomerization process and results in a higher rate of conversion. The two step chemical process is presented as follows [11]

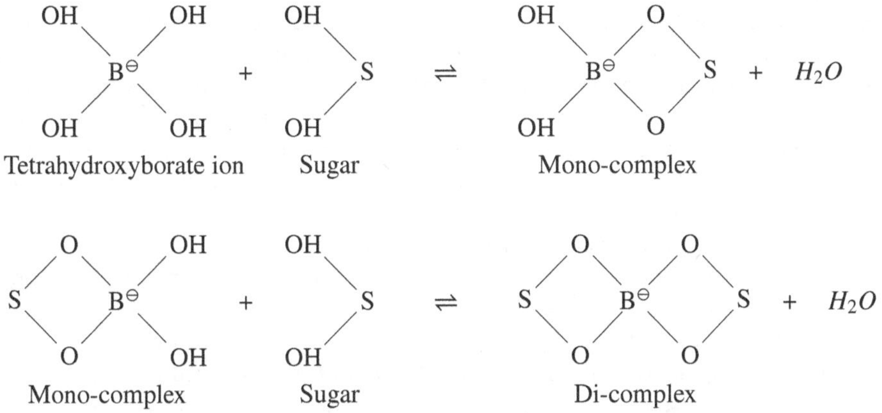

OH OH OH OH O

$B^{\ominus}$ + S ⇌ $B^{\ominus}$ S + H_2O

OH OH OH OH O

Tetrahydroxyborate ion Sugar Mono-complex

O OH OH O O

S $B^{\ominus}$ + S ⇌ S $B^{\ominus}$ S + H_2O

O OH OH O O

Mono-complex Sugar Di-complex

The described two-step reversible reaction pathway for borate ions interacting with sugar molecules may be represented as

$$
\begin{aligned}
&\Phi_1 + \Phi_2 \xrightarrow{\Lambda_1} \Phi_3 + \Phi_4 \\
&\Phi_4 + \Phi_3 \xrightarrow{\Lambda_2} \Phi_2 + \Phi_1 \\
&\Phi_3 + \Phi_2 \xrightarrow{\Lambda_3} \Phi_5 + \Phi_4 \\
&\Phi_5 + \Phi_4 \xrightarrow{\Lambda_4} \Phi_3 + \Phi_2
\end{aligned}
\tag{3.1}
$$

The concentration of various chemical compounds involved in the reactions can be represented by Φ_i, where $i = 1, 2, 3, 4, 5$. Specifically, Φ_1 represents the concentration of tetrahydroxyborate ions, Φ_2 represents the concentration of sugar molecules, Φ_3 represents the concentration of mono-complexes, Φ_4 represents the concentration of water, and Φ_5 represents the concentration of di-complexes. The rate constants of the reactions are denoted by Λ_i, where $i = 1, 2, 3, 4$.

To describe the binding of borate ions with two glucose molecules, a mathematical formulation based on the rate law is employed. This formulation takes into account various factors that influence the increase and decrease in the concentration of the chemical compounds involved in the reactions.

$$
\begin{cases}
\Phi_1'(t) = & -\Lambda_1\Phi_1(t)\Phi_2(t) + \Lambda_2\Phi_3(t)\Phi_4(t), \\
\Phi_2'(t) = & -\Lambda_1\Phi_1(t)\Phi_2(t) + \Lambda_2\Phi_3(t)\Phi_4(t) - \Lambda_3\Phi_3(t)\Phi_2(t) + \Lambda_4\Phi_4(t)\Phi_5(t), \\
\Phi_3'(t) = & \Lambda_1\Phi_1(t)\Phi_2(t) - \Lambda_2\Phi_3(t)\Phi_4(t) - \Lambda_3\Phi_3(t)\Phi_2(t) + \Lambda_4\Phi_4(t)\Phi_5(t), \\
\Phi_4'(t) = & \Lambda_1\Phi_1(t)\Phi_2(t) - \Lambda_2\Phi_3(t)\Phi_4(t) + \Lambda_3\Phi_3(t)\Phi_2(t) - \Lambda_4\Phi_4(t)\Phi_5(t), \\
\Phi_5'(t) = & \Lambda_3\Phi_3(t)\Phi_2(t) - \Lambda_4\Phi_4(t)\Phi_5(t).
\end{cases}
\tag{3.2}
$$

The two-step reversible chemical reaction can be mathematically formulated as a system of five first-order nonlinear differential equations. The first equation in the system describes the reaction between Φ_1 and Φ_2 in (3.1), resulting in the decrease of Φ_1 concentration and the formation of Φ_3 and Φ_4. Since the model considers a reversible process, the increase in Φ_1 concentration is due to the reaction between Φ_3 and Φ_4. Since Φ_1 is not involved in any other reaction, its concentration change is represented by only two nonlinear terms.

Similarly, the concentration of sugar molecules (Φ_2) is involved in both steps of the reaction, resulting in four nonlinear terms in its representation. The presence of pairs of nonlinear positive terms indicates an increase in concentration, while the negative terms indicate a decrease in concentration. The constants in the system are the rate constants that determine the formation of products from reactants.

The reversibility of the reactions ensures that all compounds involved in the mechanism remain active, and no compound becomes inactive during the process. The formation of products is determined by pairs of reactants, and the concentration of each compound exhibits nonlinearity with respect to the rate constants. The absence of linear terms in the system makes it highly nonlinear, which adds to the interest in studying its behavior, particularly its potential for exhibiting chaotic dynamics. Specifically, we are interested in discussing the system with fractional order in a discrete-time scenario.

The above system on discretization using fractional Caputo difference operator takes the following form:

$$\begin{cases} \Delta^{\vartheta}\Phi_1(n) = & -\Lambda_1\Phi_1(n-1+\vartheta)\Phi_2(n-1+\vartheta)+\Lambda_2\Phi_3(n-1+\vartheta)\Phi_4(n-1+\vartheta), \\ \Delta^{\vartheta}\Phi_2(n) = & -\Lambda_1\Phi_1(n-1+\vartheta)\Phi_2(n-1+\vartheta)+\Lambda_2\Phi_3(n-1+\vartheta)\Phi_4(n-1+\vartheta) \\ & -\Lambda_3\Phi_3(n-1+\vartheta)\Phi_2(n-1+\vartheta)+\Lambda_4\Phi_4(n-1+\vartheta)\Phi_5(n-1+\vartheta), \\ \Delta^{\vartheta}\Phi_3(n) = & \Lambda_1\Phi_1(n-1+\vartheta)\Phi_2(n-1+\vartheta)-\Lambda_2\Phi_3(n-1+\vartheta)\Phi_4(n-1+\vartheta) \\ & -\Lambda_3\Phi_3(n-1+\vartheta)\Phi_2(n-1+\vartheta)+\Lambda_4\Phi_4(n-1+\vartheta)\Phi_5(n-1+\vartheta), \\ \Delta^{\vartheta}\Phi_4(n) = & \Lambda_1\Phi_1(n-1+\vartheta)\Phi_2(n-1+\vartheta)-\Lambda_2\Phi_3(n-1+\vartheta)\Phi_4(n-1+\vartheta) \\ & +\Lambda_3\Phi_3(n-1+\vartheta)\Phi_2(n-1+\vartheta)-\Lambda_4\Phi_4(n-1+\vartheta)\Phi_5(n-1+\vartheta), \\ \Delta^{\vartheta}\Phi_5(n) = & \Lambda_3\Phi_3(n-1+\vartheta)\Phi_2(n-1+\vartheta)-\Lambda_4\Phi_4(n-1+\vartheta)\Phi_5(n-1+\vartheta). \end{cases} \tag{3.3}$$

In the proposed model, the Caputo difference operator Δ^{ϑ} is used, where $0 < \vartheta \leq 1$. The real parameters $\Lambda_i > 0$ (where i ranges from 1 to 5) and $n \in \mathbb{N}_{1-\vartheta}$ are involved in the equations. However, the original form of the model for the binding of borate ions with glucose molecules cannot be directly investigated. To overcome this challenge, we utilize the fractional sum equation method presented in [1], which leads to a numerically suitable form of the model:

$$
\begin{cases}
\Phi_1(n) = & \Phi_1(0) + \dfrac{1}{\Gamma(\vartheta)} \sum\limits_{k=1}^{n} \dfrac{\Gamma(n-k+\vartheta)}{\Gamma(n-k+1)} \left[-\Lambda_1\Phi_1(k-1)\Phi_2(k-1) + \Lambda_2\Phi_3(k-1)\Phi_4(k-1)\right], \\
\Phi_2(n) = & \Phi_2(0) + \dfrac{1}{\Gamma(\vartheta)} \sum\limits_{k=1}^{n} \dfrac{\Gamma(n-k+\vartheta)}{\Gamma(n-k+1)} \Big[-\Lambda_1\Phi_1(k-1)\Phi_2(k-1) + \Lambda_2\Phi_3(k-1)\Phi_4(k-1) \\
& -\Lambda_3\Phi_3(k-1)\Phi_2(k-1) + \Lambda_4\Phi_4(k-1)\Phi_5(k-1)\Big], \\
\Phi_3(n) = & \Phi_3(0) + \dfrac{1}{\Gamma(\vartheta)} \sum\limits_{k=1}^{n} \dfrac{\Gamma(n-k+\vartheta)}{\Gamma(n-k+1)} \Big[\Lambda_1\Phi_1(k-1)\Phi_2(k-1) - \Lambda_2\Phi_3(k-1)\Phi_4(k-1) \\
& -\Lambda_3\Phi_3(k-1)\Phi_2(k-1) + \Lambda_4\Phi_4(k-1)\Phi_5(k-1)\Big], \\
\Phi_4(n) = & \Phi_4(0) + \dfrac{1}{\Gamma(\vartheta)} \sum\limits_{k=1}^{n} \dfrac{\Gamma(n-k+\vartheta)}{\Gamma(n-k+1)} \Big[\Lambda_1\Phi_1(k-1)\Phi_2(k-1) - \Lambda_2\Phi_3(k-1)\Phi_4(k-1) \\
& +\Lambda_3\Phi_3(k-1)\Phi_2(k-1) - \Lambda_4\Phi_4(k-1)\Phi_5(k-1)\Big], \\
\Phi_5(n) = & \Phi_5(0) + \dfrac{1}{\Gamma(\vartheta)} \sum\limits_{k=1}^{n} \dfrac{\Gamma(n-k+\vartheta)}{\Gamma(n-k+1)} \left[\Lambda_3\Phi_3(k-1)\Phi_2(k-1) - \Lambda_4\Phi_4(k-1)\Phi_5(k-1)\right],
\end{cases}
\tag{3.4}
$$

where $n = 1, 2, 3 \cdots$.

3.2.2 Model of Cell-Cycle in Fission Yeast

The cell cycle is a crucial mechanism in living organisms, driving cell divisions through different phases. It varies from unicellular to multicellular organisms. Over the years, researchers have conducted mathematical analyses of the cell cycle process. For example, in 1998, Borisuk et al. performed bifurcation analysis of a mathematical model of the mitotic phase in frog eggs [6]. The influence of cyclin synthesis in the cell cycle was discussed in [7], and Pomerening et al. constructed an oscillator model of the cell cycle in 2003 [8]. The concept of bio-switching in the cell cycle and strategies for control were explored in [10]. Modeling the cell cycle is particularly important for understanding the dynamics of cancer, as it is often considered a disease of uncontrolled cell division. By constructing models, researchers aim to comprehend the regulatory mechanisms and their implications in the development of tumor cells. In single-celled organisms, the cell cycle simply involves replication, whereas in multicellular organisms, it encompasses a series of steps that lead to the development of complete organisms with internal organs, skin, and blood cells. In eukaryotic organisms with cells containing nuclei, the cell cycle consists of four stages.

$$G_1 \; phase \longrightarrow Synthesis \; phase \longrightarrow G_2 \; phase \longrightarrow M \; Phase$$

The cell cycle in eukaryotic organisms involves several stages, including the first gap phase (G1), DNA synthesis (S phase), a final stage of interphase (G2), and mitosis. During G1, the cell prepares itself for DNA synthesis. In the S phase, DNA is replicated, and in G2, a checkpoint ensures that the replicated DNA is free from damage before the cell enters mitosis, the final stage where cell division occurs. This intricate process occurs in all eukaryotic organisms. Understanding the cell cycle provides insights into the underlying biochemical processes and the biological switches that regulate the transitions between different phases.

A crucial control switch in the cell cycle is the cyclin-dependent kinase (Cdk), which stimulates the cell to progress from one phase to the next by phosphorylating specific substrates. The process itself exhibits complex behavior, involving active and inactive forms of proteins such as mitosis-promoting factor (MPF), cdc25, and Mytl. MPF is a protein complex that plays a role in promoting mitosis, while cdc25 and Mytl are protein kinases involved in the regulation of the cell cycle. The transition from G2 to the M phase (mitosis) can be described by the following equations:

$$
\begin{aligned}
\Psi_2 + \Theta_1 &\xrightarrow{\beta_1} \Psi_1 + \Psi_2 \\
\Psi_1 + \Psi_3 &\xrightarrow{\beta_2} \Psi_3 + \Theta_1 \\
\Psi_1 + \Theta_2 &\xrightarrow{\beta_3} \Psi_1 + \Psi_2 \\
\Psi_2 &\xrightarrow{\beta_4} \Theta_2 \\
\Theta_3 &\xrightarrow{\beta_5} \Psi_3 \\
\Psi_1 + \Psi_3 &\xrightarrow{\beta_6} \Psi_1 + \Theta_3
\end{aligned} \tag{3.5}
$$

where $\Psi_i, i = 1, 2, 3$ denotes the active forms of MPF, cdc25, Mytl and their respective inactive forms are denoted by $\Theta_i, i = 1, 2, 3$. The given set of reactions can be described in terms of the concentrations of the active forms of three chemical compounds involved in the cell cycle mechanism. The presence of inactive molecules is neglected in the model since they are not directly involved in the mechanism or contribute to the cell cycle process. However, the total amounts of the three molecules, namely MPF, cdc25, and Mytl, present in the solution are denoted by the constants h_1, h_2, and h_3 respectively.

The system of equations is constructed based on the molecules acting as reactants or products. In a chemical reaction where a compound reacts with another to form products, the concentration of the reactants decreases while the concentration of the products increases. For example, let's consider the modeling of the Mytl compound. It is involved in three reactions, specifically the second, fifth, and sixth reactions presented in the cell cycle equations. In the second reaction, Mytl reacts with MPF to produce the active form of Mytl and the inactive form of MPF. In this case, the concentration of Mytl remains the same when it reacts with MPF. Since there are no absolute changes in the concentration, no terms are included in the third equation of the system.

The fifth reaction represents the activation of inactive Mytl at a rate β_5. The corresponding equation is formulated by considering the difference between the total amount of Mytl and the amount of activated Mytl, as well as the product of the concentrations of Mytl and MPF. The negative sign denotes the decrease in Mytl concentration due to its reaction with MPF, resulting in the deactivation of Mytl.

$$
\begin{cases}
\xi_1'(t) = & \beta_1\xi_2(t)(h_1 - \xi_1(t)) - \beta_2\xi_1(t)\xi_3(t), \\
\xi_2'(t) = & \beta_3\xi_1(t)(h_2 - \xi_2(t)) - \beta_4\xi_2(t), \\
\xi_3'(t) = & \beta_5(h_3 - \xi_3(t)) - \beta_6\xi_1(t)\xi_3(t).
\end{cases} \tag{3.6}
$$

The concentration of the active forms of MPF, cdc25 and Mytl are denoted by $\xi_i(t), i = 1, 2, 3$ respectively and $\beta_i, i = 1, 2, 3, 4, 5, 6$ are non zero real valued rate constants of the biochemical reactions. The fractional order form derived using Caputo difference operator with $n \in \mathbb{N}_{1-\vartheta}$ is given by

$$
\begin{cases}
\Delta^{\vartheta}\xi_1(n) = & \beta_1\xi_2(n-1+\vartheta)(h_1 - \xi_1(n-1+\vartheta)) - \beta_2\xi_1(n-1+\vartheta)\xi_3(n-1+\vartheta), \\
\Delta^{\vartheta}\xi_2(n) = & \beta_3\xi_1(n-1+\vartheta)(h_2 - \xi_2(n-1+\vartheta)) - \beta_4\xi_2(n-1+\vartheta), \\
\Delta^{\vartheta}\xi_3(n) = & \beta_5(h_3 - \xi_3(n-1+\vartheta)) - \beta_6\xi_1(n-1+\vartheta)\xi_3(n-1+\vartheta).
\end{cases} \tag{3.7}
$$

The system (3.7) is converted to a numerically feasible form as follows:

$$\begin{cases} \xi_1(n) = & \xi_1(0) + \dfrac{1}{\Gamma(\vartheta)} \sum\limits_{k=1}^{n} \dfrac{\Gamma(n-k+\vartheta)}{\Gamma(n-k+1)} \left[\beta_1 \xi_2(k-1)(h_1 - \xi_1(k-1)) - \beta_2 \xi_1(k-1)\xi_3(k-1)\right], \\ \xi_2(n) = & \xi_2(0) + \dfrac{1}{\Gamma(\vartheta)} \sum\limits_{k=1}^{n} \dfrac{\Gamma(n-k+\vartheta)}{\Gamma(n-k+1)} \left[\beta_3 \xi_1(k-1)(h_2 - \xi_2(k-1)) - \beta_4 \xi_2(k-1)\right], \\ \xi_3(n) = & \xi_3(0) + \dfrac{1}{\Gamma(\vartheta)} \sum\limits_{k=1}^{n} \dfrac{\Gamma(n-k+\vartheta)}{\Gamma(n-k+1)} \left[\beta_5 (h_3 - \xi_3(k-1)) - \beta_6 \xi_1(k-1)\xi_3(k-1)\right]. \; n = 1, 2, 3, \cdots. \end{cases} \tag{3.8}$$

3.3 Chaos in Commensurate Order Systems

The analysis will be conducted for two different cases of the proposed system: a commensurate order system where the state variables have the same fractional order, and an incommensurate order system where the state variables have different fractional orders. In this section, we will focus on the commensurate order system where the state variables representing the concentration of the molecules involved in the chemical reactions are assumed to have identical fractional orders. The inclusion of fractional orders in the state variables is motivated by the need to capture specific physical properties that cannot be adequately modeled using integer order derivatives. Our expectation is that the system will exhibit chaotic behavior when the fractional order is much less than 1, making this study interesting with intriguing results on the system's behavior. Investigating the chaotic response of a biochemical system provides insights into how changes in the reaction rate constants can affect the overall mechanism of organisms. It also sheds light on how perturbations at certain stages of the reaction can have a profound impact on the entire process.

In the subsequent section, we will present the incommensurate order system to further illustrate these concepts.

3.3.1 Reaction of Borate Ions Binding with Glucose

In this section, we will explore the chaotic behavior exhibited by the mathematical model that represents the chemical reaction of borate binding with glucose molecules. To analyze and visualize the chaotic dynamics, we will utilize bifurcation diagrams with respect to the parameter Λ_3, as well as calculate the Lyapunov exponents using the Jacobian matrix method proposed in [3]. By analyzing the bifurcation diagrams and calculating the Lyapunov exponents, we can gain a deeper understanding of the chaotic nature of borate binding with glucose molecules and uncover the intricate dynamics of this biochemical reaction system.

Numerical Validation:

By considering the parameter values $\vartheta = 0.5$, $\Lambda_1 = 12$, $\Lambda_2 = 1.8$, $\Lambda_3 = 1.5$, and an initial condition of $(0.1, 0.1, 0.1, 0.1, 0.1, 0.1)$, we simulated the bifurcation diagrams for all the state variables by varying Λ_3 within the range of [0, 3]. The results of these simulations are presented in Figures 3.1, 3.2, and 3.3. From the bifurcation diagrams for each state variable, we can observe that the system, under the considered parameter values, fails to exhibit a sustained steady state when the rate constant Λ_3 of the reaction involving the formation of the di-complex is varied. For smaller values of Λ_3, the system displays complete chaotic behavior across all state variables.

It is important to note that the considered equation represents a reversible reaction with simultaneous forward and backward reactions. The concept of a steady state can be interpreted as uniform oscillations. Such steady state behavior is obtained for relatively higher values of Λ_3, specifically when $\Lambda_3 > 1.5$. However,

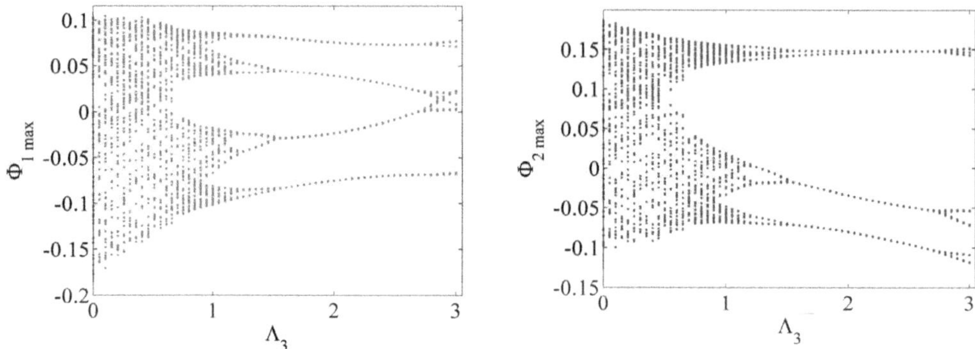

Fig. 3.1: Bifurcation diagrams for state variables Φ_1, Φ_2 of system (3.3).

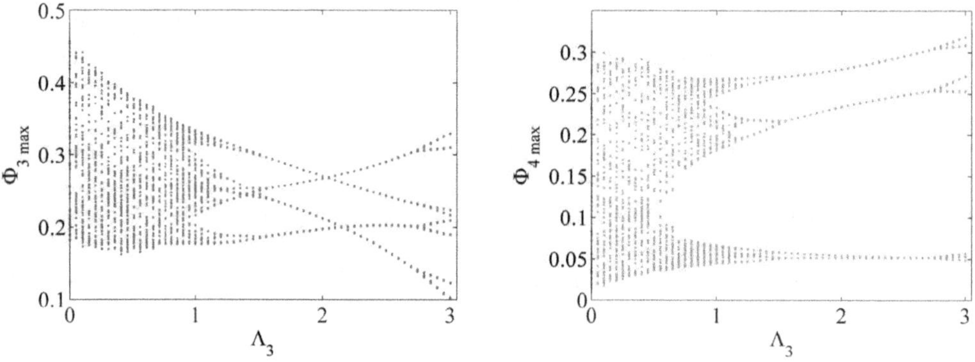

Fig. 3.2: Bifurcation diagrams for state variables Φ_3, Φ_4 of system (3.3).

beyond a certain range, the system transitions into aperiodic oscillations, resulting in chaos. The bifurcation diagrams for state variables Φ_2 and Φ_4 exhibit similar behavior. This is due to the fact that sugar molecules and water are involved in all the reversible reactions discussed. Sugar molecules combine with tetrahydroxyborate and mono-complex in two steps, while water is formed as a product in both cases. Similarly, the formation of sugar occurs in both reactions when water reacts with mono and di complexes.

Overall, the bifurcation diagrams provide valuable insights into the complex dynamics of the system under different values of the rate constant Λ_3. The observed chaotic behavior underscores the sensitivity of the biochemical reaction system to changes in this parameter and highlights the importance of understanding the underlying mechanisms governing such behavior.

We have provided time series plots and phase plane plots to demonstrate the changes in system behavior. Figures 3.4, 3.5, 3.6, 3.7, 3.8, 3.9, 3.10, 3.11 depict these plots for different values of Λ_3. In the case of biochemical systems, time series plots are often more effective in illustrating the system's behavior compared to phase plane analyses. This is because time series plots allow for a better understanding of the individual changes in the molecular concentrations. We conducted the analysis for four different values of Λ_3: 0.05, 1.25, 2, and 2.95. For $\Lambda_3 = 0.05$, the time series plot in Figure 3.4 shows random oscillations of all the state variables without any periodic behavior. The corresponding phase plane plot in Figure 3.5 illustrates the formation of a chaotic attractor.

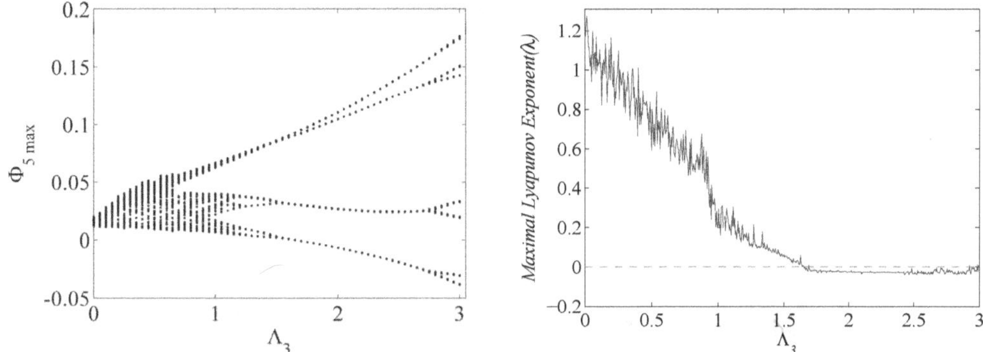

Fig. 3.3: Bifurcation diagrams for state variable Φ_5 of system (3.3) and and Lyapunov exponent.

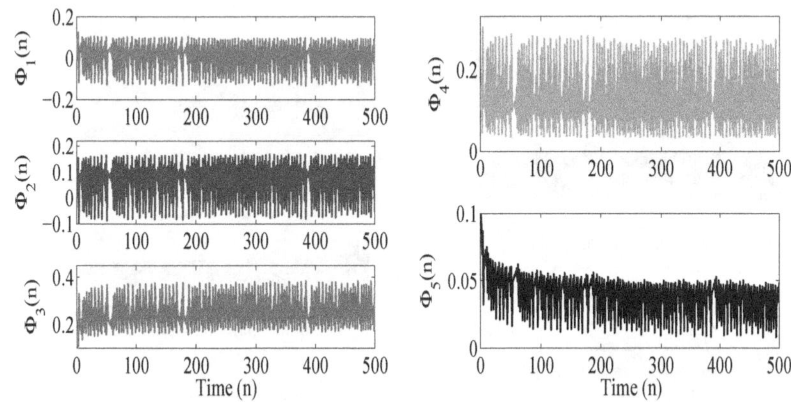

Fig. 3.4: Time series plots with respect to bifurcation diagram and Lyapunov exponent of system (3.3) for $\Lambda_3 = 0.05$.

When the value of Λ_3 is increased to 1.25, the time series plot in Figure 3.6 demonstrates that the state variables are initially perturbed but eventually settle into constant oscillations. The phase plane plot in Figure 3.7 provides additional insights into the behavior of the molecules. It is important to note that the provided time series plots and phase plane plots showcase the dynamics of the system under different values of Λ_3. These visual representations offer valuable information about the complex behavior exhibited by the biochemical system.

For higher values of the rate constant Λ_3, such as $\Lambda_3 = 2$, the system exhibits uniform oscillatory motion after an initial perturbation. This behavior is depicted in Figure 3.8, where the time series plot shows relatively stable and periodic oscillations of all the state variables. The corresponding phase plane diagram in Figure 3.9 illustrates the predictable orbit of the trajectories, indicating periodic oscillations. However, it is interesting to note that the system does not maintain periodic oscillations for all values of Λ_3 greater than 2. The bifurcation diagrams presented for the state variables demonstrate the occurrence of bifurcations and the formation of branches, indicating a chaotic response that can be observed for higher values of the rate constant Λ_3.

To further investigate the system's behavior, we study the case of $\Lambda_3 = 2.95$. As expected, the system initially exhibits perturbed behavior for a certain period of time before settling into periodic oscillations. It is noteworthy that the initial perturbation observed in this case, at $\Lambda_3 = 2.95$, is longer in duration compared

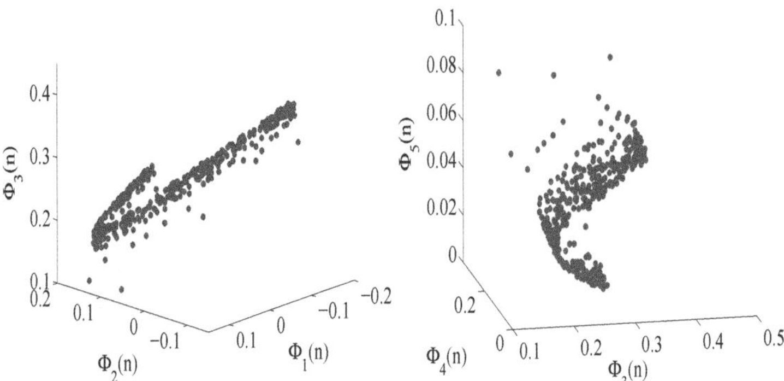

Fig. 3.5: Phase plane plots with respect to bifurcation diagrams and Lyapunov exponent of system (3.3) for $\Lambda_3 = 0.05$.

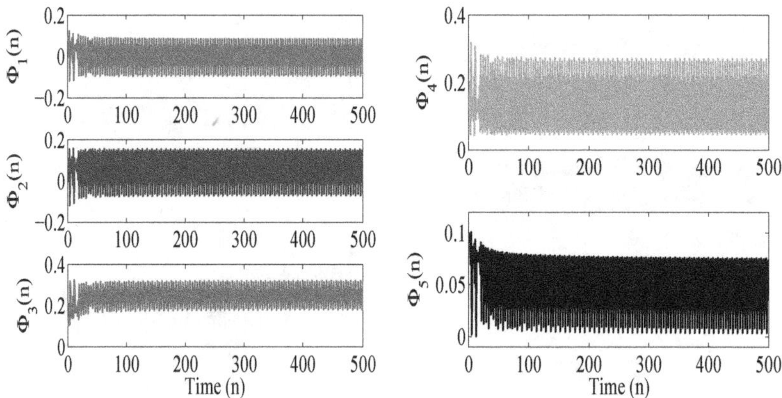

Fig. 3.6: Time series plots with respect to bifurcation diagram and Lyapunov exponent of system (3.3) for $\Lambda_3 = 1.25$.

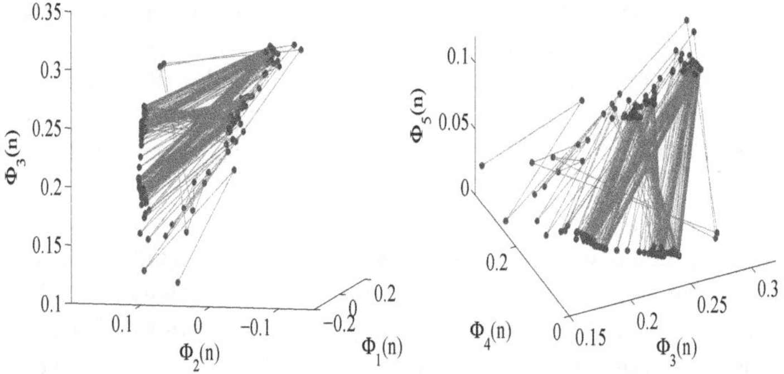

Fig. 3.7: Phase plane plots with respect to bifurcation diagram and Lyapunov exponent of system (3.3) for $\Lambda_3 = 1.25$.

to the case when $\Lambda_3 = 2$. This complexity in the behavior of the bio-chemical reaction, involving the isomerization of glucose molecules in the presence of borate ions, is highlighted by these observations.

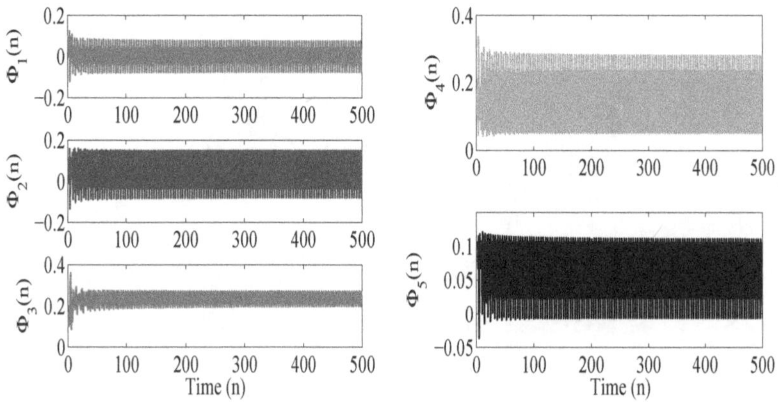

Fig. 3.8: Time series plots with respect to bifurcation diagram and Lyapunov exponent of system (3.3) for $\Lambda_3 = 2$.

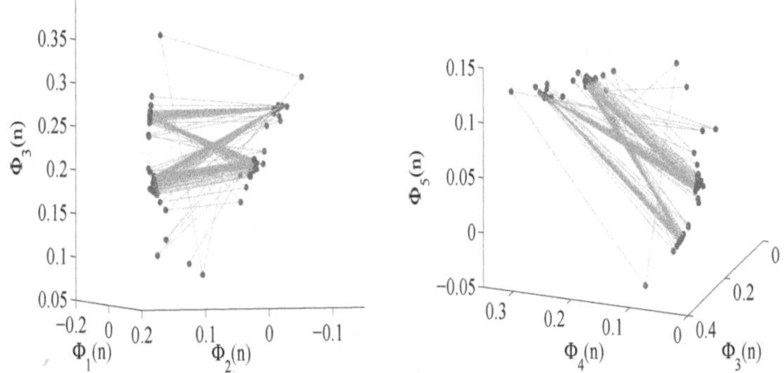

Fig. 3.9: Phase plane plots with respect to bifurcation diagram and Lyapunov exponent of system (3.3) for $\Lambda_3 = 2$.

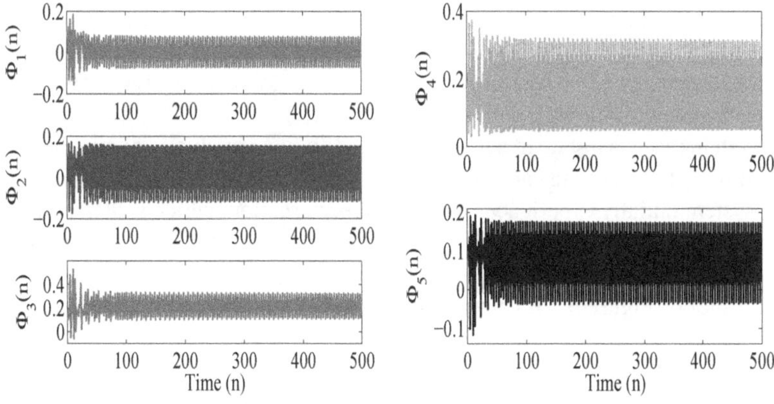

Fig. 3.10: Time series plots with respect to bifurcation diagram and Lyapunov exponent of system (3.3) for $\Lambda_3 = 2.95$.

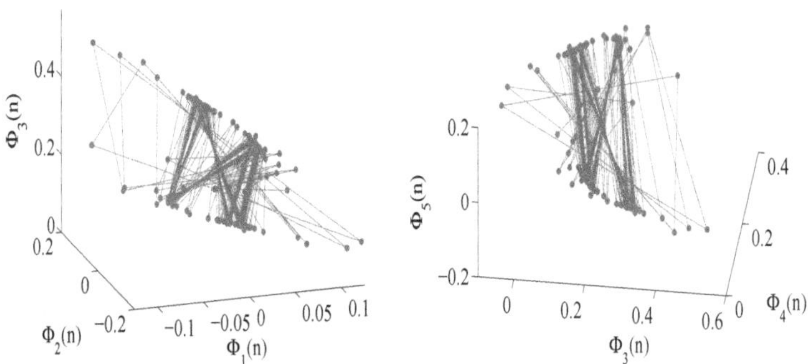

Fig. 3.11: Phase plane plots with respect to bifurcation diagram and Lyapunov exponent of system (3.3) for $\Lambda_3 = 2.95$.

3.3.2 Commensurate Order Cell Cycle Model

In this section, we assume that the state variables representing the active forms of the chemical molecules involved in the cell cycle process have identical fractional orders. We investigate the chaotic behavior of the system (3.7) by analyzing bifurcation diagrams and identifying the region of chaos through the plotting of Lyapunov exponents. The chaotic response of the system is demonstrated using an initial state of $(0.7, 0.6, 0.8)$ and the following parameter values: $\beta_1 = 2$, $\beta_3 = 0.4$, $\beta_4 = 0.55$, $\beta_5 = 0.52$, $\beta_6 = 0.7$, $h_1 = 0.5$, $h_2 = 0.7$, $h_3 = 0.9$, and $\vartheta = 0.425$. The investigation is carried out by varying the value of β_2 in the range of $(0, 2.5)$. The findings and observations are depicted in Figure 3.12. The bifurcation diagrams provide a breakdown analysis of the system's behavior, and the region of chaos can be identified by examining the corresponding Lyapunov exponents plotted against the bifurcation diagrams, thereby substantiating the presence of chaos.

To further illustrate the results, time series plots are presented in Figures 3.13 and 3.14 for different values of β_2. Additionally, 3-D phase plane plots are provided in conjunction with the time series plots for the case of $\beta_2 = 2.3$. The considered system is composed of three discrete-time fractional-order systems, making the analysis particularly suitable and convenient for obtaining phase plane plots. Time series plots are generated for each state variable at four different values of $\beta_2 : 0.5, 1.5, 2, \textit{and } 2.15$. Examining the bifurcation diagrams for the minimum rate constant value of β_2, which is involved in the reaction that leads to the deactivation of MPF, reveals that the system exhibits stable behavior for all three active forms of the molecules. However, as the rate of MPF deactivation increases, the system undergoes significant changes in its behavior, which are captured in the time series plots.

For $\beta_2 = 0.5$, the system reaches a state of stability with oscillations gradually diminishing and eventually becoming constant. At $\beta_2 = 1.5$, the state variables exhibit uniform oscillations as time progresses. When β_2 is increased to 2, the system initially experiences a period of perturbation, which is then controlled, resulting in constant oscillations. This glimpse of aperiodic oscillations indicates that higher values of β_2 will likely lead to chaos. To closely examine these changes, we further increase β_2 to 2.15, and the resulting behavior can be observed in the last plot of Figure 3.13. Unstable oscillations eventually cease and transition to uniform oscillations.

For $\beta_2 = 2.3$, continuous and unstable oscillations are displayed by the system, as shown in Figure 3.14. The chaotic attractor is depicted as a 3-D phase plane trajectory, providing a visual representation of the chaotic nature of the system.

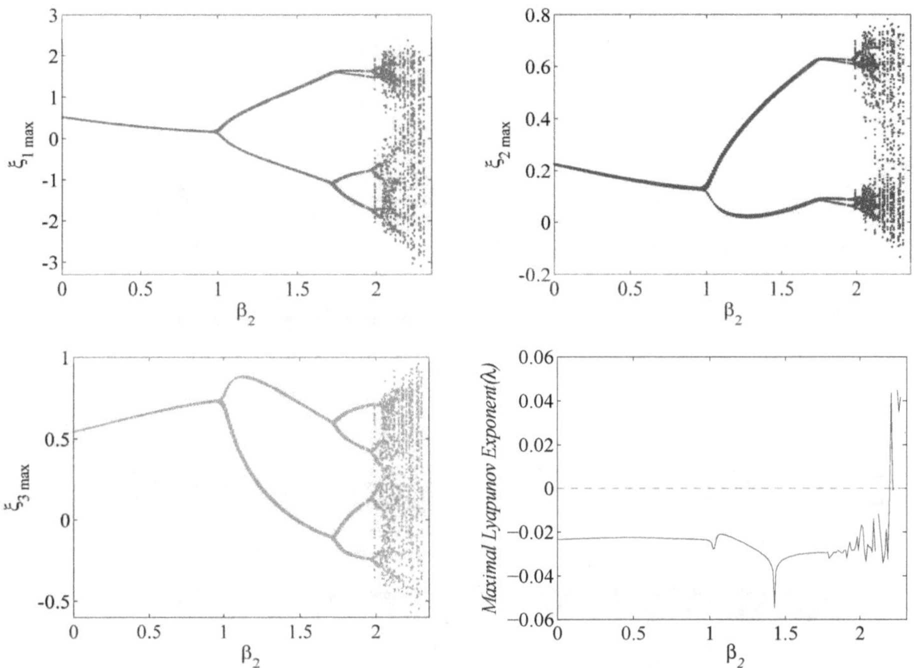

Fig. 3.12: Bifurcation diagrams and Lyapunov exponents of system (3.7).

3.4 Chaos in Incommensurate Order Systems

The investigation of biological systems using mathematical models has demonstrated its significance and relevance in accurately representing the phenomena being studied. Additionally, natural events often exhibit nonlinear behavior, making it essential to consider the diverse characteristics of different state variables when modeling such systems. Therefore, this section of the chapter focuses on exploring the behavior of the considered bio-chemical systems by introducing non-identical fractional orders for the state variables. By incorporating different fractional orders, we introduce additional parameters for investigation, enhancing our ability to delve deeper into the proposed phenomenon and gain a comprehensive understanding of the system's behavior. This approach allows us to capture the complexities and intricacies of the biological processes under investigation more effectively.

3.4.1 Isomerization of Glucose Molecules with Borate Ions

This section of the chapter aims to compare the results obtained from systems with identical fractional orders and systems with non-identical fractional orders. While constructing systems with non-identical fractional orders introduces additional complexity by increasing the number of parameters to be considered, the significance and enriched dynamics it offers make it worth exploring. Although studying systems with non-identical fractional orders may pose a greater burden due to the increased parameters involved compared to systems with identical fractional orders, the insights gained from this analysis are valuable. It allows for a more comprehensive understanding of the dynamics and behavior exhibited by these bio-chemical systems,

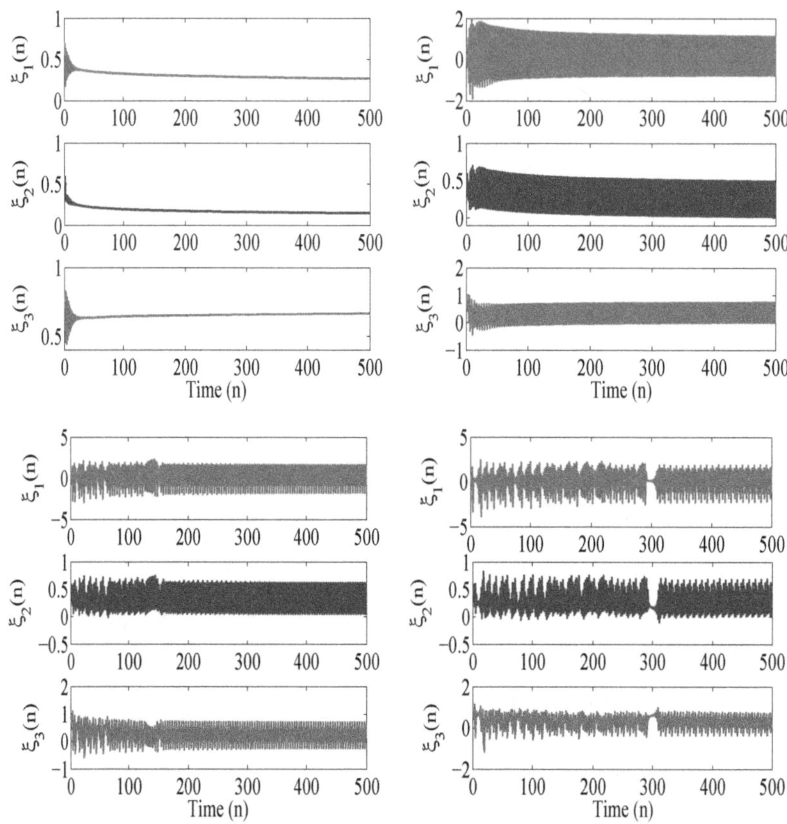

Fig. 3.13: Time series plots with respect to bifurcation diagram and Lyapunov exponent of system (3.7) in Figure 3.12 for $\beta_2 = 0.5, 1.5, 2, 2.15$.

enhancing our knowledge of their intricate dynamics. The system (3.3) with incommensurate order takes the following form as

$$\begin{cases} \Delta^{\vartheta_1}\Phi_1(n) = & -\Lambda_1\Phi_1(n-1+\vartheta_1)\Phi_2(n-1+\vartheta_1)+\Lambda_2\Phi_3(n-1+\vartheta_1)\Phi_4(n-1+\vartheta_1), \\ \Delta^{\vartheta_2}\Phi_2(n) = & -\Lambda_1\Phi_1(n-1+\vartheta_2)\Phi_2(n-1+\vartheta_2)+\Lambda_2\Phi_3(n-1+\vartheta_2)\Phi_4(n-1+\vartheta_2) \\ & -\Lambda_3\Phi_3(n-1+\vartheta_2)\Phi_2(n-1+\vartheta_2)+\Lambda_4\Phi_4(n-1+\vartheta_2)\Phi_5(n-1+\vartheta_2), \\ \Delta^{\vartheta_3}\Phi_3(n) = & \Lambda_1\Phi_1(n-1+\vartheta_3)\Phi_2(n-1+\vartheta_3)-\Lambda_2\Phi_3(n-1+\vartheta_3)\Phi_4(n-1+\vartheta_3) \\ & -\Lambda_3\Phi_3(n-1+\vartheta_3)\Phi_2(n-1+\vartheta_3)+\Lambda_4\Phi_4(n-1+\vartheta_3)\Phi_5(n-1+\vartheta_3), \\ \Delta^{\vartheta_4}\Phi_4(n) = & \Lambda_1\Phi_1(n-1+\vartheta_4)\Phi_2(n-1+\vartheta_4)-\Lambda_2\Phi_3(n-1+\vartheta_4)\Phi_4(n-1+\vartheta_4) \\ & +\Lambda_3\Phi_3(n-1+\vartheta_4)\Phi_2(n-1+\vartheta_4)-\Lambda_4\Phi_4(n-1+\vartheta_4)\Phi_5(n-1+\vartheta_4), \\ \Delta^{\vartheta_5}\Phi_5(n) = & \Lambda_3\Phi_3(n-1+\vartheta_5)\Phi_2(n-1+\vartheta_5)-\Lambda_4\Phi_4(n-1+\vartheta_5)\Phi_5(n-1+\vartheta_5), \end{cases} \tag{3.9}$$

where $0 < \vartheta_i < 1, i = 1,2,3,4,5$ are the fractional orders of the state variables $\Phi_i(n), i = 1,2,3,4,5$ respectively and $n \in \mathbb{N}_{1-\vartheta_i}, i = 1,2,3,4,5$. In the case of the model describing the isomerization of glucose molecules in the presence of borate ions, the number of parameters involved in the analysis increases from

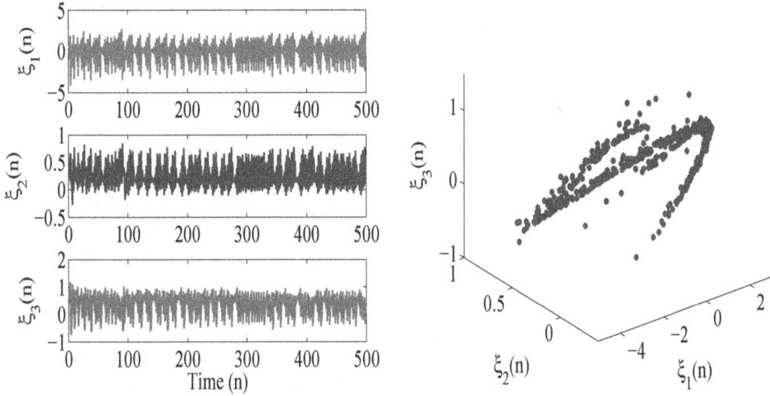

Fig. 3.14: Time series and phase plane plot with respect to bifurcation diagram and Lyapunov exponent of system (3.7) in Figure 3.12 for $\beta_2 = 2.3$.

five including the fractional order, in the commensurate case to nine parameters in the incommensurate order system, considering five different fractional orders for the five state variables and four mechanism-based parameters. As a result, the study becomes more complex, with each parameter playing a distinct role in the dynamics of the entire system. For the chaos analysis, we consider the following parameter values: $\Lambda_1 = 12$, $\Lambda_2 = 1.8$, $\Lambda_4 = 1.75$, and fractional orders $\vartheta_1 = 0.5$, $\vartheta_2 = 0.47$, $\vartheta_3 = 0.49$, $\vartheta_4 = 0.46$, $\vartheta_5 = 0.48$. The initial states are set to $(0.1, 0.1, 0.1, 0.1, 0.1, 0.1)$, and the parameter Λ_3 is varied from 0 to 3 to investigate the dynamic changes occurring in the chemical reaction system (3.3). The resulting bifurcation diagrams are presented in Figure 3.15.

Although the overall appearance of the bifurcation diagrams may seem similar to those obtained for the commensurate order systems, a closer examination reveals changes in the amplitude of the diagrams. This emphasizes the importance of Figures 3.18, 3.19, 3.20, 3.21, 3.22, 3.23, 3.24, 3.25 for illustrating the dynamics. The time series plots provide insight into the transition of the variables' states, from aperiodic and chaotic behavior for lower values of Λ_3 to controlled perturbation during the initial time and eventually uniform oscillations as time progresses. The subsequent transition from uniform oscillations to chaos can be observed from the bifurcation diagrams for the state variables, along with the Lyapunov exponent, which remains close to the zero line and takes positive values as Λ_3 approaches 3.

In this study, we further explore the system by simultaneously varying the fractional order at three different values, while keeping the other parameters fixed. Specifically, we investigate the bifurcation behavior at $\vartheta_1 = 0.45, 0.65, 0.85$. This analysis highlights the additional degree of freedom that incommensurate order systems offer, allowing us to uncover fascinating features of the system.

Figures 3.26 and 3.27 present a sequence of bifurcation diagrams simulated with Λ_3 as the bifurcation parameter for the three different fractional order values. When the state variable $\Phi_1(n)$ has a fractional order of 0.45, the system exhibits complete chaos, as evident from the 2D surfaces depicted in Figures 3.26 and 3.27. As we increase the value of ϑ_1 to 0.65, the system bifurcates and displays chaos over a smaller range. Further increasing ϑ_1 to 0.85 results in chaotic behavior occurring only within a limited range of Λ_3. These findings demonstrate the importance of studying the impact of fractional orders on different state variables, as it provides valuable insights for developing control strategies for the system.

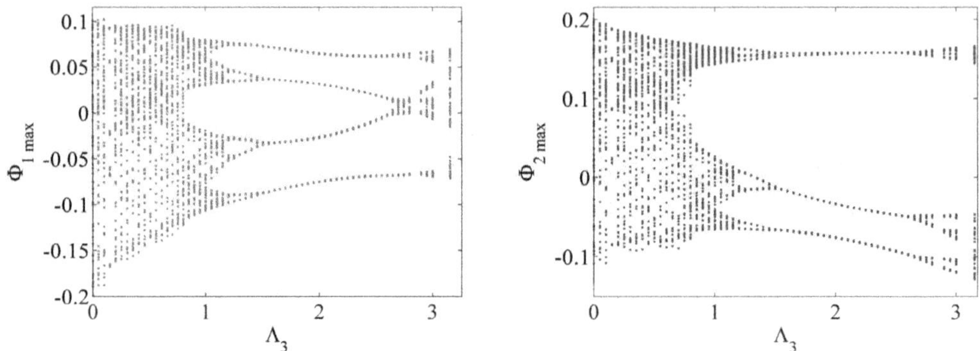

Fig. 3.15: Bifurcation diagrams for state variables Φ_1, Φ_2 of system (3.9).

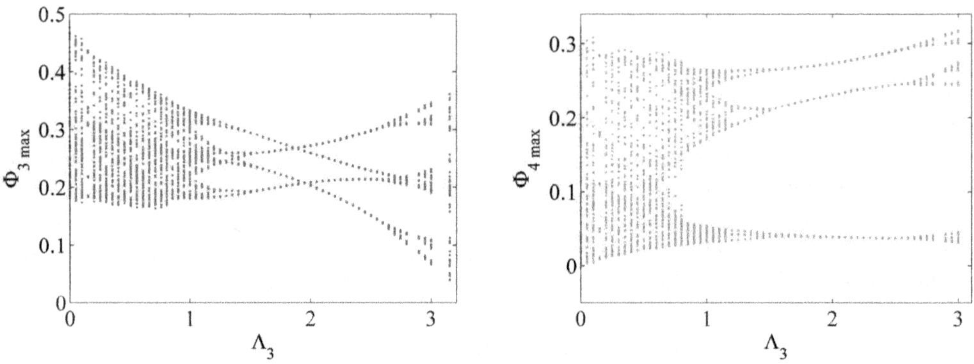

Fig. 3.16: Bifurcation diagrams for state variables Φ_3, Φ_4 of system (3.9).

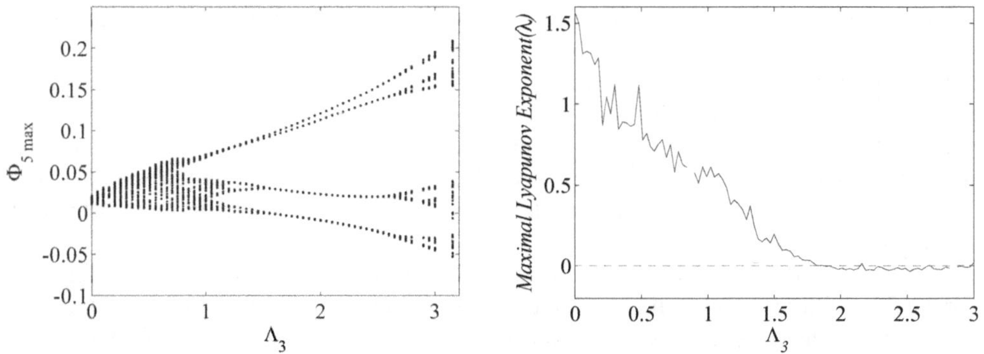

Fig. 3.17: Bifurcation diagrams for state variable Φ_5 of system (3.9) and Lyapunov exponents.

3.4.2 Cell Cycle Model in Eukaryotes

In order to highlight the distinct behavior and importance of the state variables in the context of normal cell cycles in eukaryotes, this section focuses on assigning different fractional orders to each of these variables.

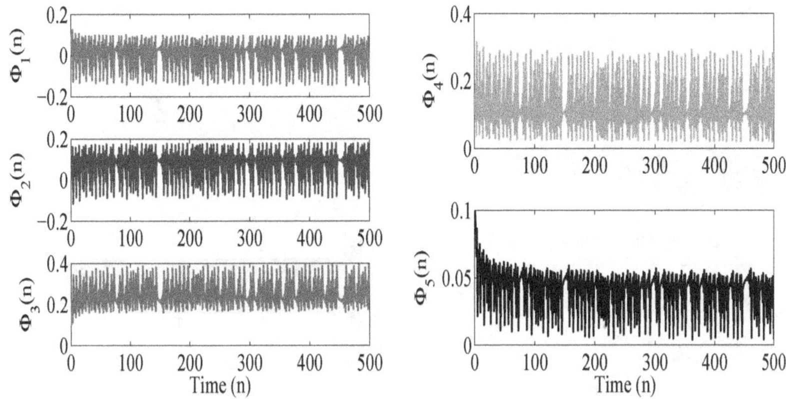

Fig. 3.18: Time series plots with respect to bifurcation diagram and Lyapunov exponent of system (3.9) for $\Lambda_3 = 0.05$.

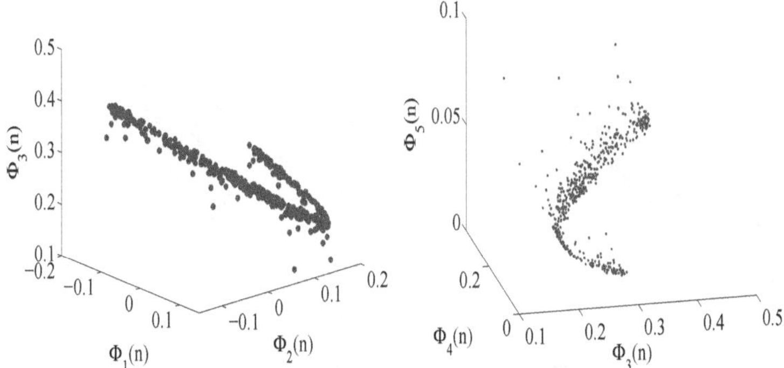

Fig. 3.19: Phase plane plots with respect to bifurcation diagram and Lyapunov exponent of system (3.9) for $\Lambda_3 = 0.05$.

By introducing non-identical fractional orders, we aim to investigate and analyze the system numerically, emphasizing the individuality and unique contributions of each state variable. This approach allows us to gain a deeper understanding of the dynamics and characteristics of the system in the context of cell cycle processes. The system (3.7) with incommensurate order is given by

$$\begin{cases} \Delta^{\vartheta_1}\xi_1(n) = & \beta_1\xi_2(n-1+\vartheta_1)(h_1-\xi_1(n-1+\vartheta_1)) - \beta_2\xi_1(n-1+\vartheta_1)\xi_3(n-1+\vartheta_1), \\ \Delta^{\vartheta_2}\xi_2(n) = & \beta_3\xi_1(n-1+\vartheta_2)(h_2-\xi_2(n-1+\vartheta_2)) - \beta_4\xi_2(n-1+\vartheta_2), \\ \Delta^{\vartheta_3}\xi_3(n) = & \beta_5(h_3-\xi_3(n-1+\vartheta_3)) - \beta_6\xi_1(n-1+\vartheta_3)\xi_3(n-1+\vartheta_3), \end{cases} \tag{3.10}$$

where $0 < \vartheta_i < 1, i = 1,2,3$ are the fractional orders of the state variables $\xi_i, i = 1,2,3$ respectively and $n \in \mathbb{N}_{1-\vartheta_i}, i = 1,2,3$.

Numerical simulations:

The simulations presented in Figure 3.28 illustrate the occurrence of chaos in the system with non-identical fractional orders. The parameters used for these simulations are $\vartheta_1 = 0.44$, $\vartheta_2 = 0.45$, $\vartheta_3 = 0.46$, $\beta_1 = 2$, $\beta_3 = 0.4$, $\beta_4 = 0.6$, $\beta_5 = 0.52$, $\beta_6 = 0.7$, $h_1 = 0.5$, $h_2 = 0.7$, $h_3 = 0.9$, and the initial states are set

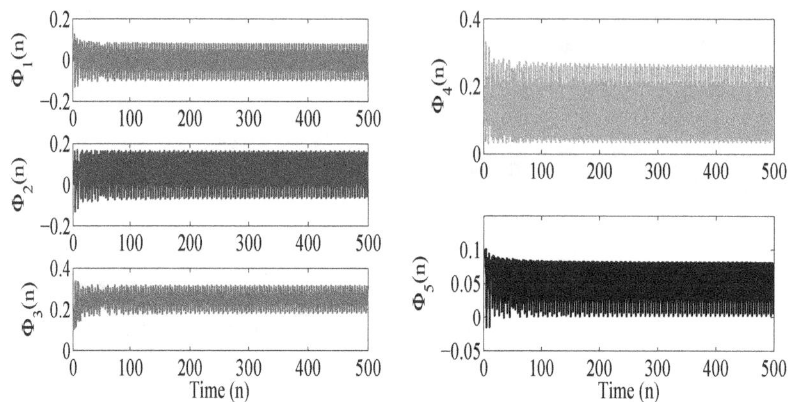

Fig. 3.20: Time series plots with respect to bifurcation diagram and Lyapunov exponent of system (3.9) for $\Lambda_3 = 1.25$.

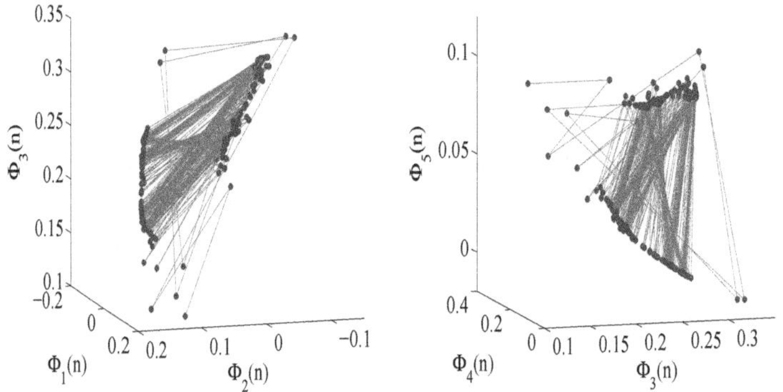

Fig. 3.21: Phase plane plots with respect to bifurcation diagram and Lyapunov exponent of system (3.9) for $\Lambda_3 = 1.25$.

to $(0.7, 0.5, 0.8)$. The value of β_2 is varied continuously between 0 and 2.5. In the analysis, we initially consider a small difference in fractional orders of ± 0.01 from the value of 0.45. It becomes evident that even a small difference in fractional orders leads to significant variations in the system's behavior. The bifurcation diagrams for this system have the same parameter values as the commensurate order case, except for β_4 which is increased to 0.6. Despite this small change, the system exhibits enhanced chaos.

To compare the chaotic behavior between the commensurate order system (Figure 3.12) and the incommensurate order system (Figure 3.28), the Lyapunov exponents are analyzed. It can be observed that in the incommensurate order system, the Lyapunov exponents consistently lie above the zero line for a wider range of β_2 values, indicating an expanded region of chaos. To provide a detailed understanding of the system's dynamics at different β_2 values, time series plots are shown in Figure 3.29. These plots illustrate the transitions and changes in the system's behavior. Additionally, for $\beta_2 = 2.4$, the chaotic response is represented through a phase plane plot and accompanying time series plots in Figure 3.30.

Overall, the study of the system with non-identical fractional orders reveals enhanced chaos and provides insights into the system's behavior under varying parameter values.

Figure 3.31 emphasizes the significance of constructing models with different fractional orders and highlights the advantages they bring for analysis. It demonstrates the behavior of the system when the

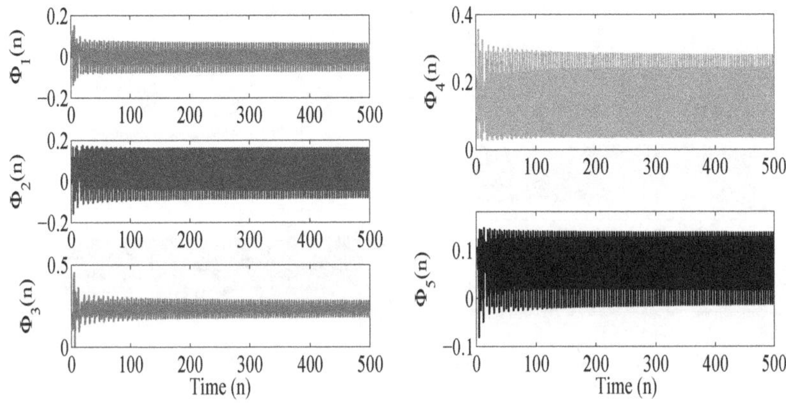

Fig. 3.22: Time series plots with respect to bifurcation diagram and Lyapunov exponent of system (3.9) for $\Lambda_3 = 2.25$.

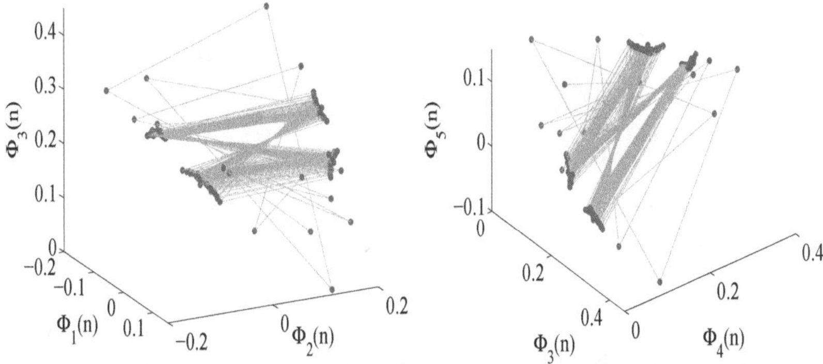

Fig. 3.23: Phase plane plots with respect to bifurcation diagram and Lyapunov exponent of system (3.9) for $\Lambda_3 = 2.25$.

fractional order of the state variable $\xi_1(n)$ is close to 0.25 and 0.75. In both cases, the system undergoes bifurcations, but no chaotic response is observed. However, when ϑ_1 is set to 0.5, the system exhibits chaotic behavior. A similar analysis can be performed by varying each of the fractional orders ϑ_i individually while keeping the other parameters fixed. By varying β_2, the complex behavior of the system can be understood more comprehensively. This type of analysis enables a thorough understanding of the behavior of individual variables and their interactions within the system.

3.5 Synchronization of Bio-Chemical Models

In this section, nonlinear control functions with feedback gains are used to synchronise the master and slave biochemical models. We will review the stability criteria for discrete fractional equations from the following theorems before moving on to the synchronisation procedure.

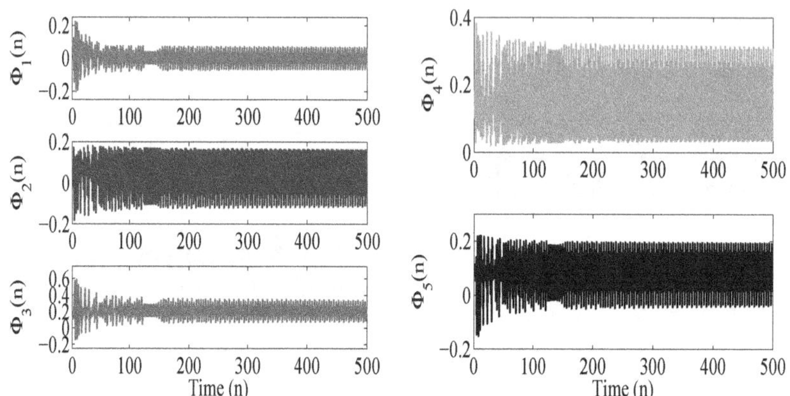

Fig. 3.24: Time series plots with respect to bifurcation diagram and Lyapunov exponent of system (3.9) for $\Lambda_3 = 3$.

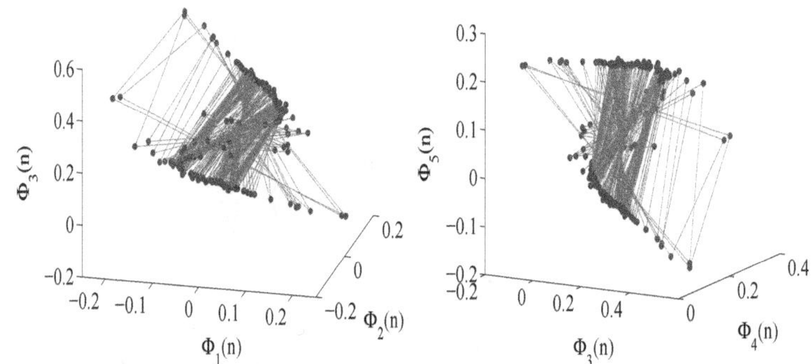

Fig. 3.25: Phase plane plots with respect to bifurcation diagram and Lyapunov exponent of system (3.9) for $\Lambda_3 = 3$.

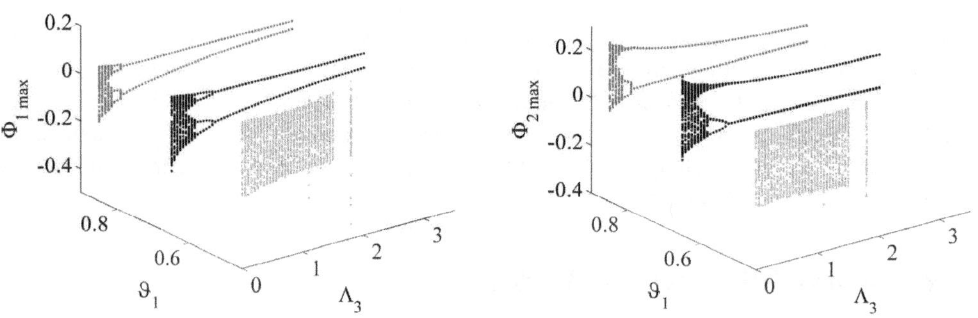

Fig. 3.26: Sequence of bifurcation diagrams for state variables Φ_1, Φ_2 of system (3.9).

Theorem 3.1 *[2]Let $\vartheta \in (0, 1)$ and $\Psi \in \mathbb{R}^{k\times k}$. Let*

$$B^{\vartheta} = \left\{ \lambda \in \mathbb{C} : |\lambda| < \left(2\cos\left(\frac{|arg\lambda| - \phi}{2-\vartheta}\right)\right)^{\vartheta} \text{ and } |arg\lambda| > \frac{\vartheta\pi}{2} \right\}.$$

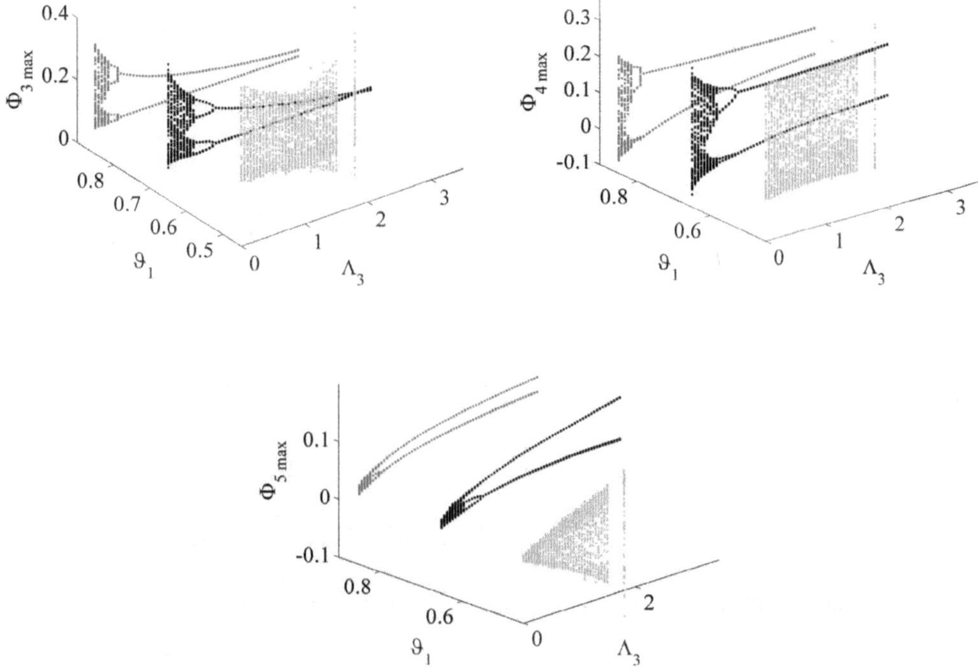

Fig. 3.27: Sequence of bifurcation diagrams for state variables Φ_3, Φ_4, Φ_5 of system (3.9).

If all the eigenvalues of Ψ *lie in* B^{ϑ}*, then the system* (6.1) *is asymptotically stable. The system* (6.1) *is unstable if* $\lambda \in \mathbb{C}\backslash cl(B^{\vartheta})$ *for an eigenvalue of* Ψ*.*

Theorem 3.2 *[9] Let* $\alpha_{\kappa} = \dfrac{\omega_{\kappa}}{\beta_{\kappa}} \in (0,1)$*, where* $\kappa = 1, 2, 3, \cdots, m$*,* $\omega_{\kappa}, \beta_{\kappa} > 0$ *are positive integers, the LCM of the denominators is* Φ*. The trivial solution of the system* (6.1) *with initial condition* $x(0) = x_0$ *is asymptotically stable if the equation*

$$det\left(diag\left(\eta^{\Phi\alpha_1}, \eta^{\Phi\alpha_2}, \ldots, \eta^{\Phi\alpha_m}\right) - (1 - \eta^{\Phi}J)\right) = 0, \tag{3.11}$$

has all the roots inside the set $\mathbb{C}\backslash B_1^{\vartheta}$*, where* $\vartheta = \dfrac{1}{\Phi}$ *and*

$$B_1^{\vartheta} = \left\{\lambda \in \mathbb{C} : |\lambda| \le \left(2\cos\frac{|arg\lambda|}{\vartheta}\right)^{\vartheta} \text{ and } |arg\lambda| \le \frac{\vartheta\pi}{2}\right\}.$$

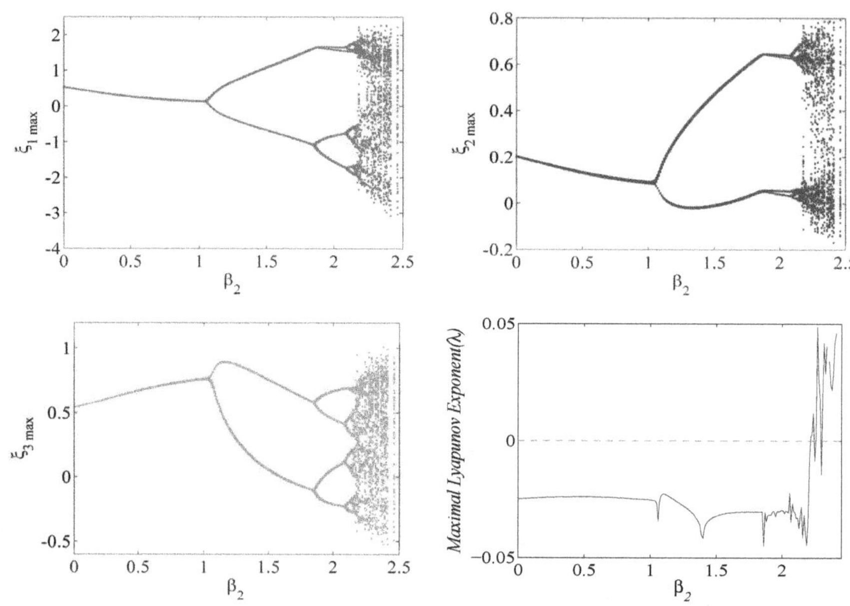

Fig. 3.28: Bifurcation diagrams and Lyapunov exponents of system (3.10).

3.5.1 Discrete fractional model of cell cycle in fission yeast

In order to achieve the synchronization of the cell-cycle models, we first describe the master and slave version of the (3.10) as follows.

$$\begin{cases} \Delta^{\vartheta_1}(\xi_1)_m(n) = & \beta_1(\xi_2)_m(n-1+\vartheta_1)(h_1-(\xi_1)_m(n-1+\vartheta_1))-\beta_2(\xi_1)_m(n-1+\vartheta_1)(\xi_3)_m(n-1+\vartheta_1), \\ \Delta^{\vartheta_2}(\xi_2)_m(n) = & \beta_3(\xi_1)_m(n-1+\vartheta_2)(h_2-(\xi_2)_m(n-1+\vartheta_2))-\beta_4(\xi_2)_m(n-1+\vartheta_2), \\ \Delta^{\vartheta_3}(\xi_3)_m(n) = & \beta_5(h_3-(\xi_3)_m(n-1+\vartheta_3))-\beta_6(\xi_1)_m(n-1+\vartheta_3)(\xi_3)_m(n-1+\vartheta_3), \end{cases} \tag{3.12}$$

and

$$\begin{cases} \Delta^{\vartheta_1}(\xi_1)_s(n) = & \beta_1(\xi_2)_s(n-1+\vartheta_1)(h_1-(\xi_1)_s(n-1+\vartheta_1))-\beta_2(\xi_1)_s(n-1+\vartheta_1)(\xi_3)_s(n-1+\vartheta_1)+H_1(n-1+\vartheta_1), \\ \Delta^{\vartheta_2}(\xi_2)_s(n) = & \beta_3(\xi_1)_s(n-1+\vartheta_2)(h_2-(\xi_2)_s(n-1+\vartheta_2))-\beta_4(\xi_2)_s(n-1+\vartheta_2)+H_2(n-1+\vartheta_2), \\ \Delta^{\vartheta_3}(\xi_3)_s(n) = & \beta_5(h_3-(\xi_3)_s(n-1+\vartheta_3))-\beta_6(\xi_1)_s(n-1+\vartheta_3)(\xi_3)_s(n-1+\vartheta_3)+H_3(n-1+\vartheta_3), \end{cases} \tag{3.13}$$

where $H_1(n)$, $H_2(n)$, $H_3(n)$ are the nonlinear control functions and $((\xi_1)_m(0),(\xi_2)_m(0),(\xi_3)_m(0)) = (\alpha_1,\alpha_2,\alpha_3)$ $((\xi_1)_s(0),(\xi_2)_s(0),(\xi_3)_s(0)) = (\alpha_4,\alpha_5,\alpha_6)$ are the initial conditions with $\alpha_i, i = 1,2,3,4,5,6$ are real constants.

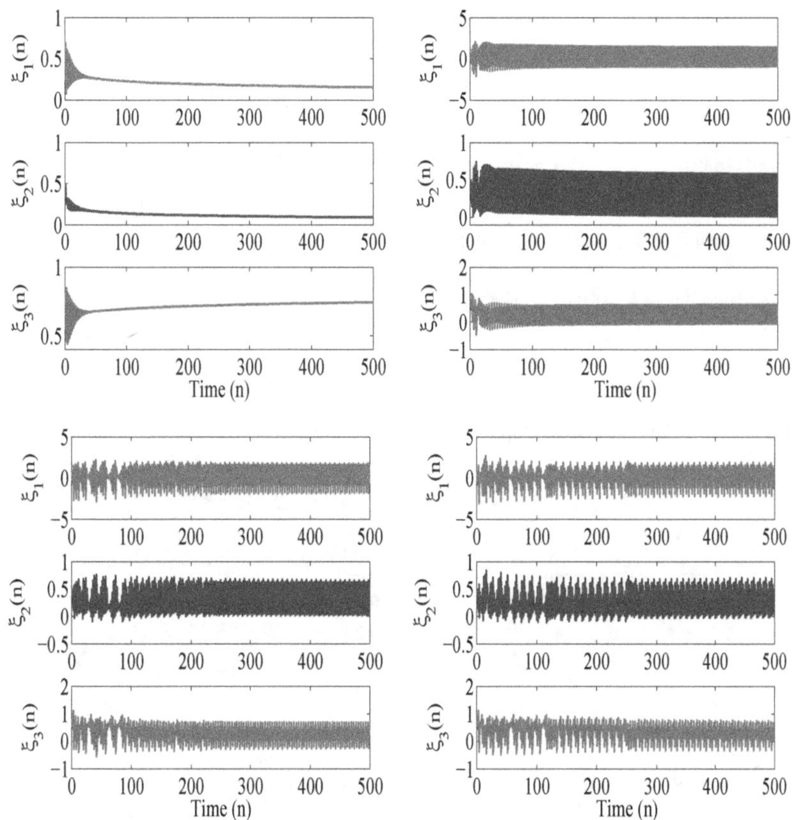

Fig. 3.29: Time series plots with respect to bifurcation diagram and Lyapunov exponent of system (3.10) in Figure 3.28 for $\beta_2 = 0.75, 1.75, 2.15, 2.25$.

Theorem 3.3 *The master system* (3.12) *and the slave system* (3.13) *of the cell cycle of fission yeast model* (3.7) *is synchronized if the control functions are defined as*

$$\begin{aligned} H_1(n) =&\beta_1[(\xi_2)_s(n)E_{\xi_1}(n) + (\xi_1)_m(n)E_{\xi_2}(n)] + \beta_2[(\xi_3)_s(n)E_{\xi_1}(n) + (\xi_1)_m(n)E_{\xi_3}(n)] - K_1E_{\xi_1}(n) \\ H_2(n) =&\beta_3[(\xi_2)_s(n)E_{\xi_1}(n) + (\xi_1)_m(n)E_{\xi_2}] - K_2E_{\xi_2}(n) \\ H_3(n) =&\beta_6[(\xi_3)_s(n)E_{\xi_1}(n) + (\xi_1)_m(n)E_{\xi_3}] - K_3E_{\xi_3}(n), \end{aligned} \tag{3.14}$$

where $E_{\xi_i}(n), i = 1, 2, 3$ *are the error states of synchronization with feedback gains* K_1, K_2, K_3.

Proof We shall first derive the error states of the synchronization process. The error states are defined by

$$\begin{bmatrix} E_{\xi_1} \\ E_{\xi_2} \\ E_{\xi_3} \end{bmatrix} = \begin{bmatrix} (\xi_1)_s \\ (\xi_2)_s \\ (\xi_3)_s \end{bmatrix} - \begin{bmatrix} (\xi_1)_m \\ (\xi_2)_m \\ (\xi_3)_m \end{bmatrix}. \tag{3.15}$$

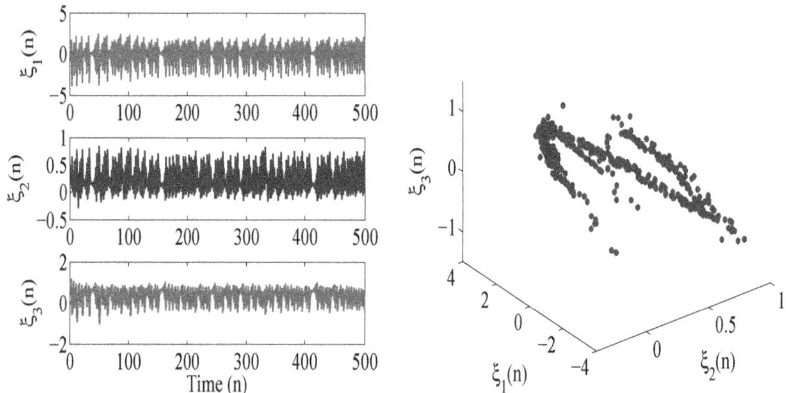

Fig. 3.30: Time series and phase plane plot with respect to bifurcation diagram and Lyapunov exponent of system (3.10) in Figure 3.28 for $\beta_2 = 2.4$.

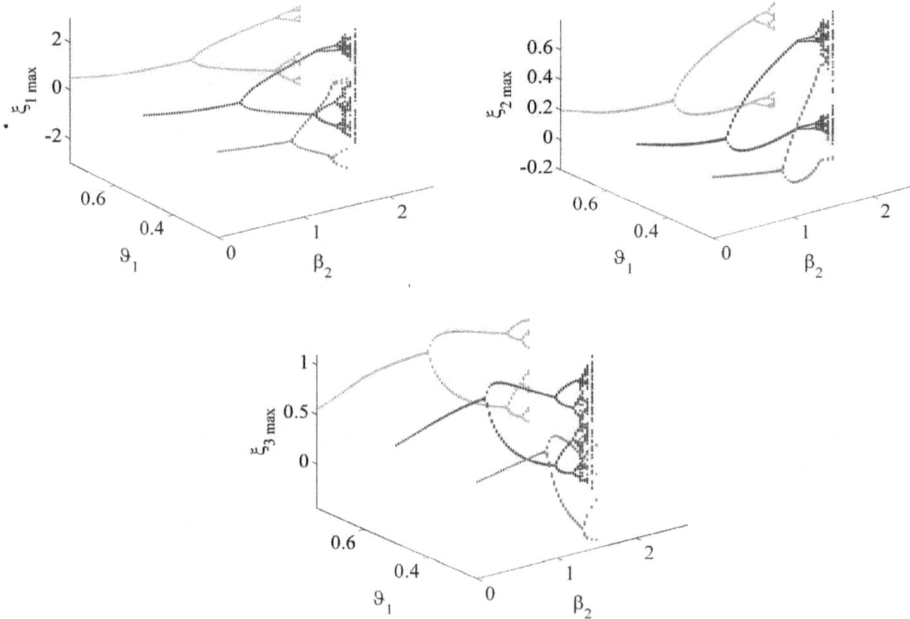

Fig. 3.31: Sequence of bifurcation diagrams and Lyapunov exponents of system (3.10).

Then the corresponding error dynamical system can be obtained as

$$\begin{cases}\Delta^{\vartheta_1} E_{\xi_1}(n) = & \beta_1 h_1 E_{\xi_2}(n-1+\vartheta_1) - K_1 E_{\xi_1}(n-1+\vartheta_1)\\ \Delta^{\vartheta_2} E_{\xi_2}(n) = & \beta_3 h_2 E_{\xi_1}(n-1+\vartheta_2) - [\beta_4 + K_2] E_{\xi_2}(n-1+\vartheta_2)\\ \Delta^{\vartheta_3} E_{\xi_3}(n) = & -[\beta_5 + K_3] E_{\xi_3}(n-1+\vartheta_3).\end{cases} \tag{3.16}$$

The asymptotic stability of the error states of the system (3.16) will be proved numerically in two cases. Case 1 is for commensurate order and Case 2 is for incommensurate order.

Case-1:

We shall prove the asymptotic stability for the following parameters $\beta_1 = 0.6, \beta_3 = 0.2, \beta_4 = 0.55, \beta_5 = 0.25$, $h_1 = 0.35$, $h_2 = 0.6$, $K_1 = 1.65$, $K_2 = 0.95$, $K_3 = 1.5$ with fractional order $\vartheta_1 = \vartheta_2 = \vartheta_3 = 0.85$. Eigenvalues corresponding to the dynamical system (3.16) are $\eta_1 = -1.750570499$, $\eta_2 = -1.399429501$, $\eta_3 = -1.75$. From the condition, the value of $\frac{\vartheta\pi}{2} < 1.571428$ for all $\vartheta \in (0,1)$. We know that the argument of any negative real number is equal to π and argument of any positive number is zero. Thus, we have

$$argument(-\eta_i) = \pi, \tag{3.17}$$

for all $i = 1, 2, 3$. Thus, the second inequality from Theorem 3.1,

$$|arg\eta| > \frac{\vartheta\pi}{2}$$

is true. In case of inequality,

$$\left(2\cos\left(\frac{|arg\eta| - \pi}{2-\vartheta}\right)\right)^{\vartheta} = 2^{\vartheta}.$$

It is clear that the system (3.16) converges asymptotically to the origin if

$$|\eta_i| < 2^{\vartheta}, i = 1, 2, 3. \tag{3.18}$$

The condition is true for all the eigenvalues with $\vartheta = 0.85$. Thus, error states for the commensurate fractional order converges asymptotically to origin. Simulations in Figure 3.32 support the analytically obtained result.

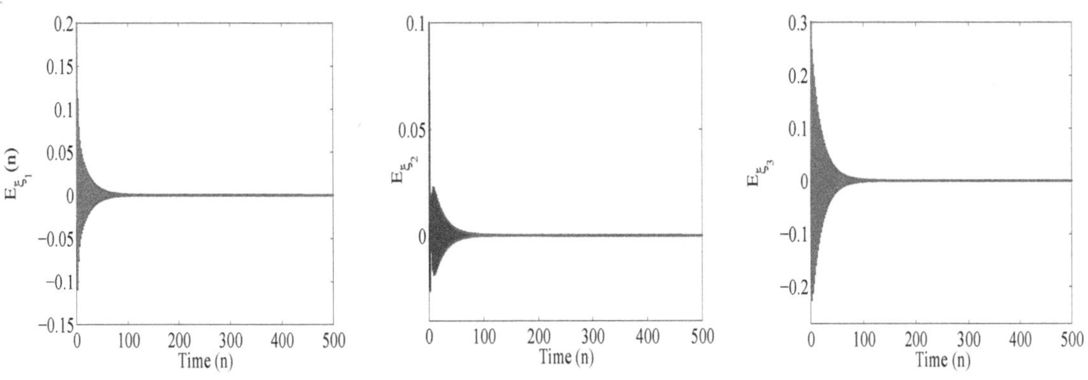

Fig. 3.32: Asymptotic stability of the error states in (3.16).

Case-2:

We will demonstrate the asymptotic stability of the non-identical order error states in equation (3.16) using Theorem 3.2. For the given parameters $\beta_1 = 0.6$, $\beta_3 = 0.2$, $\beta_4 = 0.55$, $\beta_5 = 0.15$, $h_1 = 0.35$, $h_2 = 0.6$, $K_1 = 1.65$, $K_2 = 0.95$, $K_3 = 1.25$, and non-identical fractional orders $\vartheta_1 = 0.9$, $\vartheta_2 = 0.8$, $\vartheta_3 = 0.7$, the stability analysis is performed. Theorem 3.2 is applied to prove the asymptotic stability of the error states. Using the equation (3.11), we arrive at the following:

$$\begin{aligned} &\eta^{24} + 1.4\eta^{17} + 1.5\eta^{16} + 1.65\eta^{15} + 2.10\eta^{9} \\ &+ 2.310\eta^{8} + 2.4498\eta^{7} + 3.42972 = 0 \end{aligned} \tag{3.19}$$

Corresponding eigenvalues are presented in Table 3.1. It is clear that

Table 3.1: Roots of the equation (3.19).

η_1 = 0.94533+0.45524 i	η_{13} = -0.40189+0.97144 i
η_2 = 0.23347+1.02293 i	η_{14}= -0.80956+0.67902 i
η_3 = -0.65419+0.82032 i	η_{15} = -0.97122+0.40243 i
η_4 = -1.04924	η_{16} = -1.05663
η_5 = -0.65419-0.82032 i	η_{17}= -0.97122-0.40243 i
η_6 = 0.23347-1.02293 i	η_{18} = -0.80956-0.67902 i
η_7 = 0.94533-0.45524 i	η_{19} = -0.40189-0.97144 i
η_8 = 0.99212+0.36497 i	η_{20} = -0.18398-1.04049 i
η_9 = 0.97219+0.39879 i	η_{21} = 0.40323-0.97085 i
η_{10} = 0.52742+0.91562 i	η_{22} = 0.52742-0.91562 i
η_{11} = 0.40323+0.97085 i	η_{23} = 0.97219-0.39879 i
η_{12} = -0.18398+1.04049 i	η_{24}= 0.99212-0.36497 i

$|arg\eta_l| > \dfrac{\pi}{20}$ for $l \in \{1, 2, 3, 4, 5, 6, 7, 8, 9, 10, 11, 12, 13, 14, 15, 16, 17, 18, 19, 20, 21, 22, 23, 24\}$. From Theorem 3.2 $\eta_l \in \mathbb{C}\backslash B_1^{\vartheta}$ for $1 \le l \le 24$, ensures the asymptotic stability of the system at the origin. Visual analysis for the results are presented for each error state in Figure 3.33.

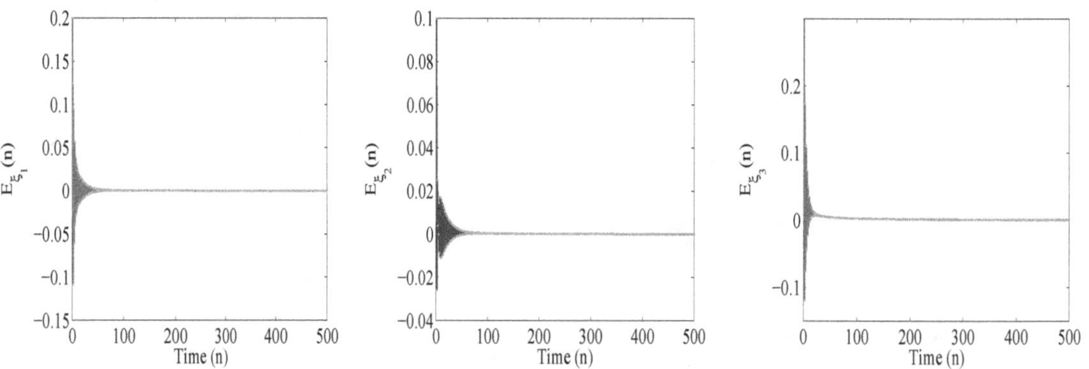

Fig. 3.33: Asymptotic stability of the error states in (3.16) for non-identical fractional orders.

3.5.2 Discrete fractional isomerization of glucose molecules with borate ions

The aim of this section is to successfully synchronise systems of glucose molecule isomerization with master and slave versions. Master version of (3.3) is given as

$$\begin{cases} \Delta^{\vartheta}(\Phi_1)_m(n) = & -\Lambda_1(\Phi_1)_m(n-1+\vartheta)(\Phi_2)_m(n-1+\vartheta)+\Lambda_2(\Phi_3)_m(n-1+\vartheta)(\Phi_4)_m(n-1+\vartheta),\\ \Delta^{\vartheta}(\Phi_2)_m(n) = & -\Lambda_1(\Phi_1)_m(n-1+\vartheta)(\Phi_2)_m(n-1+\vartheta)+\Lambda_2(\Phi_3)_m(n-1+\vartheta)(\Phi_4)_m(n-1+\vartheta)\\ & -\Lambda_3(\Phi_3)_m(n-1+\vartheta)(\Phi_2)_m(n-1+\vartheta)+\Lambda_4(\Phi_4)_m(n-1+\vartheta)(\Phi_5)_m(n-1+\vartheta),\\ \Delta^{\vartheta}(\Phi_3)_m(n) = & \Lambda_1(\Phi_1)_m(n-1+\vartheta)(\Phi_2)_m(n-1+\vartheta)-\Lambda_2(\Phi_3)_m(n-1+\vartheta)(\Phi_4)_m(n-1+\vartheta)\\ & -\Lambda_3(\Phi_3)_m(n-1+\vartheta)(\Phi_2)_m(n-1+\vartheta)+\Lambda_4(\Phi_4)_m(n-1+\vartheta)(\Phi_5)_m(n-1+\vartheta),\\ \Delta^{\vartheta}(\Phi_4)_m(n) = & \Lambda_1(\Phi_1)_m(n-1+\vartheta)(\Phi_2)_m(n-1+\vartheta)-\Lambda_2(\Phi_3)_m(n-1+\vartheta)(\Phi_4)_m(n-1+\vartheta)\\ & +\Lambda_3(\Phi_3)_m(n-1+\vartheta)(\Phi_2)_m(n-1+\vartheta)-\Lambda_4(\Phi_4)_m(n-1+\vartheta)(\Phi_5)_m(n-1+\vartheta),\\ \Delta^{\vartheta}(\Phi_5)_m(n) = & \Lambda_3(\Phi_3)_m(n-1+\vartheta)(\Phi_2)_m(n-1+\vartheta)-\Lambda_4(\Phi_4)_m(n-1+\vartheta)(\Phi_5)_m(n-1+\vartheta), \end{cases} \tag{3.20}$$

and the slave version is

$$\begin{cases} \Delta^{\vartheta}(\Phi_1)_s(n) = & -\Lambda_1(\Phi_1)_s(n-1+\vartheta)(\Phi_2)_s(n-1+\vartheta)+\Lambda_2(\Phi_3)_s(n-1+\vartheta)(\Phi_4)_s(n-1+\vartheta)+H_1(n-1+\vartheta),\\ \Delta^{\vartheta}(\Phi_2)_s(n) = & -\Lambda_1(\Phi_1)_s(n-1+\vartheta)(\Phi_2)_s(n-1+\vartheta)+\Lambda_2(\Phi_3)_s(n-1+\vartheta)(\Phi_4)_s(n-1+\vartheta)\\ & -\Lambda_3(\Phi_3)_s(n-1+\vartheta)(\Phi_2)_s(n-1+\vartheta)+\Lambda_4(\Phi_4)_s(n-1+\vartheta)(\Phi_5)_s(n-1+\vartheta)+H_2(n-1+\vartheta),\\ \Delta^{\vartheta}(\Phi_3)_s(n) = & \Lambda_1(\Phi_1)_s(n-1+\vartheta)(\Phi_2)_s(n-1+\vartheta)-\Lambda_2(\Phi_3)_s(n-1+\vartheta)(\Phi_4)_s(n-1+\vartheta)\\ & -\Lambda_3(\Phi_3)_s(n-1+\vartheta)(\Phi_2)_s(n-1+\vartheta)+\Lambda_4(\Phi_4)_s(n-1+\vartheta)(\Phi_5)_s(n-1+\vartheta)+H_3(n-1+\vartheta),\\ \Delta^{\vartheta}(\Phi_4)_s(n) = & \Lambda_1(\Phi_1)_s(n-1+\vartheta)(\Phi_2)_s(n-1+\vartheta)-\Lambda_2(\Phi_3)_s(n-1+\vartheta)(\Phi_4)_s(n-1+\vartheta)\\ & +\Lambda_3(\Phi_3)_s(n-1+\vartheta)(\Phi_2)_s(n-1+\vartheta)-\Lambda_4(\Phi_4)_s(n-1+\vartheta)(\Phi_5)_s(n-1+\vartheta)+H_4(n-1+\vartheta),\\ \Delta^{\vartheta}(\Phi_5)_s(n) = & \Lambda_3(\Phi_3)_s(n-1+\vartheta)(\Phi_2)_s(n-1+\vartheta)-\Lambda_4(\Phi_4)_s(n-1+\vartheta)(\Phi_5)_s(n-1+\vartheta)+H_5(n-1+\vartheta), \end{cases} \tag{3.21}$$

where $H_i, i = 1, 2, 3, 4, 5$ are the nonlinear control functions, initial states of the master and slave version are denoted by $(\Phi_i)_m(0) = \alpha_i$ and $(\Phi_i)_s(0) = \gamma_i$ with $\alpha_i, \gamma(i) \in \mathbb{R}$, for $i = 1, 2, 3, 4, 5$.

Theorem 3.4 *The master system* (3.20) *and the slave system* (3.21) *of the isomerization of glucose with borate ions* (3.3) *is synchronized if the control functions are defined as*

$$
\begin{aligned}
H_1(n) =& \Lambda_1(\Phi_1)_s(n)(\Phi_2)_s(n) - \Lambda_2(\Phi_3)_s(n)(\Phi_4)_s(n) - K_1 E_{\Phi_1} \\
& - \Lambda_1(\Phi_1)_m(n)(\Phi_2)_m(n) + \Lambda_2(\Phi_3)_m(n)(\Phi_4)_m(n) \\
H_2(n) =& \Lambda_1(\Phi_1)_s(n)(\Phi_2)_s(n) - \Lambda_2(\Phi_3)_s(n)(\Phi_4)_s(n) \\
& + \Lambda_3(\Phi_3)_s(n)(\Phi_2)_s(n) - \Lambda_4(\Phi_4)_s(n)(\Phi_5)_s(n) - K_2 E_{\Phi_2} \\
& - \Lambda_1(\Phi_1)_m(n)(\Phi_2)_m(n) + \Lambda_2(\Phi_3)_m(n)(\Phi_4)_m(n) \\
& - \Lambda_3(\Phi_3)_m(n)(\Phi_2)_m(n) + \Lambda_4(\Phi_4)_m(n)(\Phi_5)_m(n) \\
H_3(n) =& - \Lambda_1(\Phi_1)_s(n)(\Phi_2)_s(n) + \Lambda_2(\Phi_3)_s(n)(\Phi_4)_s(n) \\
& + \Lambda_3(\Phi_3)_s(n)(\Phi_2)_s(n) - \Lambda_4(\Phi_4)_s(n)(\Phi_5)_s(n) - K_3 E_{\Phi_3} \\
& + \Lambda_1(\Phi_1)_m(n)(\Phi_2)_m(n) - \Lambda_2(\Phi_3)_m(n)(\Phi_4)_m(n) \\
& - \Lambda_3(\Phi_3)_m(n)(\Phi_2)_m(n) + \Lambda_4(\Phi_4)_m(n)(\Phi_5)_m(n) \\
H_4(n) =& - \Lambda_1(\Phi_1)_s(n)(\Phi_2)_s(n) + \Lambda_2(\Phi_3)_s(n)(\Phi_4)_s(n) \\
& - \Lambda_3(\Phi_3)_s(n)(\Phi_2)_s(n) + \Lambda_4(\Phi_4)_s(n)(\Phi_5)_s(n) - K_4 E_{\Phi_4} \\
& + \Lambda_1(\Phi_1)_m(n)(\Phi_2)_m(n) - \Lambda_2(\Phi_3)_m(n)(\Phi_4)_m(n) \\
& + \Lambda_3(\Phi_3)_m(n)(\Phi_2)_m(n) - \Lambda_4(\Phi_4)_m(n)(\Phi_5)_m(n) \\
H_5(n) =& - \Lambda_3(\Phi_3)_s(n)(\Phi_2)_s(n) + \Lambda_4(\Phi_4)_s(n)(\Phi_5)_s(n) - K_5 E_{\Phi_5} \\
& + \Lambda_3(\Phi_3)_m(n)(\Phi_2)_m(n) - \Lambda_4(\Phi_4)_m(n)(\Phi_5)_m(n)
\end{aligned}
\tag{3.22}
$$

where $E_{\xi_i}(n), i = 1, 2, 3, 4, 5$ are the error states of synchronization with K_i being the feedback gains.

Proof Let the error states be defined as

$$
\begin{pmatrix} E_{\Phi_1} \\ E_{\Phi_2} \\ E_{\Phi_3} \\ E_{\Phi_4} \\ E_{\Phi_5} \end{pmatrix} = \begin{pmatrix} (\Phi_1)_s \\ (\Phi_2)_s \\ (\Phi_3)_s \\ (\Phi_4)_s \\ (\Phi_5)_s \end{pmatrix} - \begin{pmatrix} (\Phi_1)_m \\ (\Phi_2)_m \\ (\Phi_3)_m \\ (\Phi_4)_m \\ (\Phi_5)_m \end{pmatrix}. \tag{3.23}
$$

The error dynamical system employing nonlinear controllers defined in Theorem 3.22 is obtained as

$$
\begin{cases}
\Delta^{\vartheta} E_{\Phi_1}(n) = & -K_1 E_{\Phi_1}(n - 1 + \vartheta), \\
\Delta^{\vartheta} E_{\Phi_2}(n) = & -K_2 E_{\Phi_2}(n - 1 + \vartheta), \\
\Delta^{\vartheta} E_{\Phi_3}(n) = & -K_3 E_{\Phi_3}(n - 1 + \vartheta), \\
\Delta^{\vartheta} E_{\Phi_4}(n) = & -K_4 E_{\Phi_4}(n - 1 + \vartheta), \\
\Delta^{\vartheta} E_{\Phi_5}(n) = & -K_5 E_{\Phi_5}(n - 1 + \vartheta),
\end{cases}
\tag{3.24}
$$

To analytically verify the asymptotic stability of the error state system (3.24), we can utilize the conditions stated in Theorem 3.1. The system (3.24) is a linear commensurate order system with the origin as its unique equilibrium state. Our goal is to demonstrate the convergence of the state variables to the origin. The eigenvalues of the system can be obtained from the Jacobian matrix, and they are denoted as $\eta_i = -K_i$, where $i = 1, 2, 3, 4, 5, 6$. In this case, all the eigenvalues are negative and real. According to the condition stated in the theorem, we have $\frac{\vartheta\pi}{2} < 1.571428$ for all $\vartheta \in (0, 1)$. It is worth noting that the argument of any negative real number is equal to π, while the argument of any positive number is zero. Based on these conditions and observations, we can proceed with the proof of the asymptotic stability of the error state system (3.24). Thus, we have

$$argument(-K_i) = \pi, \tag{3.25}$$

for all $i = 1, 2, 3, 4, 5$. Thus, irrespective of the value of the feedback gain $K_i, i = 1, 2, 3, 4, 5$, the second inequality of the condition (7.5)

$$|arg\eta| > \frac{\vartheta\pi}{2}$$

is true. In case of the first inequality, from (3.25)

$$\left(2\cos\left(\frac{|arg\eta| - \pi}{2 - \vartheta}\right)\right)^{\vartheta} = 2^{\vartheta}.$$

It is clear that the system (3.24) converges asymptotically to the origin if

$$|K_i| < 2^{\vartheta}, i = 1, 2, 3, 4, 5. \tag{3.26}$$

Numerical verification of the asymptotic stability of the error states is performed for the given parameter values: $K_1 = 1.62$, $K_2 = 1.52$, $K_3 = 1.54$, $K_4 = 1.63$, $K_5 = 1.55$, and fractional order $\vartheta = 0.8$. The fractional exponent $2^{0.8} = 1.74110$ is greater than the absolute values of all the eigenvalues obtained for the specified parameter values. This observation confirms that the error states in system (3.24) asymptotically converge to the trivial equilibrium state. The simulation results are illustrated in Figure 3.34.

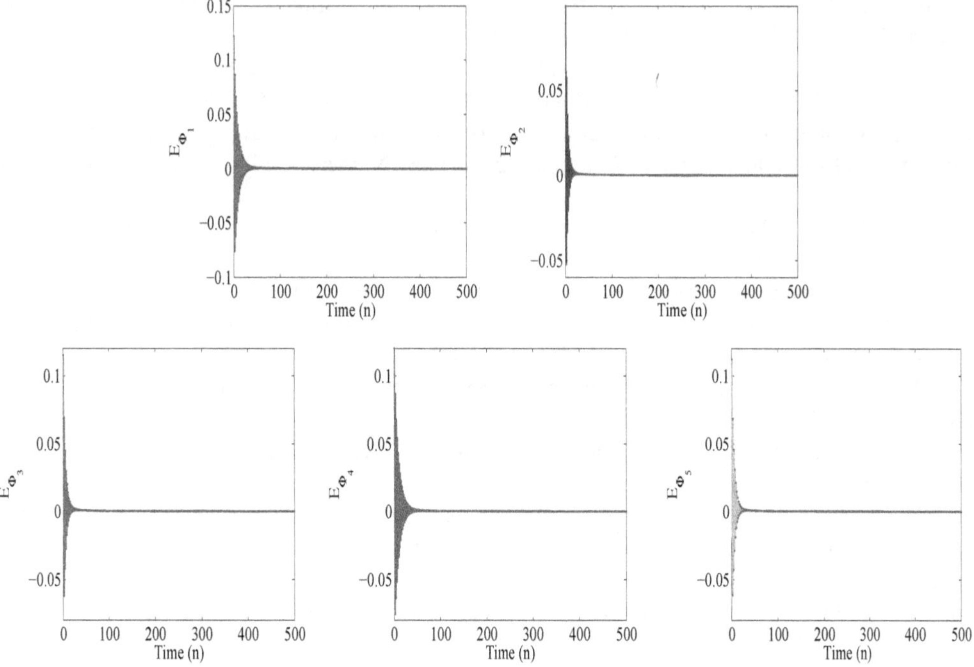

Fig. 3.34: Asymptotic stability of the error states in (3.24).

3.6 Summary

This chapter delves into the fascinating field of modeling bio-chemical reactions, which are fundamental to life and scientific advancements. It explores two different systems with significant implications for human beings. The first system focuses on the development of efficient jet fuels from bio-fuels, while the second system examines the bio-switching reactions involved in cell division, which is essential for reproduction. The mathematical models in this chapter utilize the discrete-time Caputo fractional operator, and the existence of chaos is investigated through simulations that include bifurcation diagrams and Lyapunov exponents. These analyses provide evidence for the presence of chaos in the systems. In the second part of the chapter, the study extends to systems with commensurate order, where the state variables are assigned different fractional orders. This approach allows for a more detailed exploration of the system's behavior. Bifurcation diagrams are presented in distinct segments, and time series plots are provided at selected points to facilitate a deeper understanding of the constructed models. The advantages of constructing incommensurate order models are highlighted through a sequence of bifurcation diagrams obtained by simultaneously varying two parameters. This analysis showcases the complexity of system behavior and the emergence of chaotic dynamics. Such insights are crucial for developing effective control strategies that can contribute to the production of bio-fuels in the first model and ensure the smooth functioning of cell cycles through bio-switching.

References

[1] Adel Ouannas, Amina Aicha Khennaoui, Shaher Momani, Giuseppe Grassi and Viet-Thanh Pham. Chaos and control of a three-dimensional fractional order discrete-time system with no equilibrium and its synchronization. AIP Advances 10, no. 4 (2020): 045310.

[2] Jan Cermak, Istvan Gyori and Ludek Nechvatal. On explicit stability conditions for a linear fractional difference system. Fractional Calculus and Applied Analysis 18 (2015): 651–672.

[3] Guo-Cheng Wu and Dumitru Baleanu. Jacobian matrix algorithm for Lyapunov exponents of the discrete fractional maps. Communications in Nonlinear Science and Numerical Simulation 22, no. 1-3 (2015): 95–100.

[4] Richard J. Field and F. W. Schneider Oscillating chemical reactions and nonlinear dynamics. Journal of Chemical Education 66, no. 3 (1989): 195.

[5] Richard J. Field. Chaos in chemistry and biochemistry. World Scientific (1993).

[6] Mark T. Borisuk and John J. Tyson. Bifurcation analysis of a model of mitotic control in frog eggs. Journal of Theoretical Biology 195, no. 1 (1998): 69–85.

[7] Andrew W. Murray and Marc W. Kirschner. Cyclin synthesis drives the early embryonic cell cycle. Nature 339, no. 6222 (1989): 275–280.

[8] Joseph R. Pomerening, Eduardo D. Sontag and James E. Ferrell. Building a cell cycle oscillator: Hysteresis and bistability in the activation of Cdc2. Nature Cell Biology 5, no. 4 (2003): 346–351.

[9] Mohd Taib Shatnawi, Noureddine Djenina, Adel Ouannas, Iqbal M. Batiha and Giuseppe Grassi. Novel convenient conditions for the stability of nonlinear incommensurate fractional-order difference systems. Alexandria Engineering Journal 61, no. 2 (2022): 1655–1663.

[10] Thron, C. D. Bistable biochemical switching and the control of the events of the cell cycle. Oncogene 15, no. 3 (1997): 317–325.

[11] Boris M. Slepchenko and Mark Terasaki. Cyclin aggregation and robustness of bio-switching. Molecular Biology of the Cell 14, no. 11 (2003): 4695–4706.

[12] Wei Wang, Ashutosh Mittal, Heidi Pilath, Xiaowen Chen, Melvin P. Tucker and David K. Johnson. Simultaneous upgrading of biomass-derived sugars to HMF/furfural via enzymatically isomerized ketose intermediates. Biotechnology for Biofuels 12, no. 1 (2019): 1–9.

[13] Suqin Hu, Zhaofu Zhang, Yinxi Zhou, Buxing Han, Honglei Fan, Wenjing Li, Jinliang Song and Ye Xie. Conversion of fructose to 5-hydroxymethylfurfural using ionic liquids prepared from renewable materials. Green Chemistry 10, no. 12 (2008): 1280–1283.

[14] Yoshiyuki Takasaki. Studies on sugar-isomerizing enzymes: Effect of borate on glucose-fructose isomerization catalyzed by glucose isomerase. Agricultural and Biological Chemistry 35, no. 9 (1971): 1371–1375.

[15] Mahar, T. J. and B. J. Matkowsky. A model biochemical reaction exhibiting secondary bifurcation. SIAM Journal on Applied Mathematics, 32, no. 2 (1977): 394–404.

[16] Yilei Tang and Weinian Zhang. Bogdanov-Takens bifurcation of a polynomial differential system in biochemical reaction. Computers & Mathematics with Applications 48, no. 5-6 (2004): 869–883.

[17] Hongwei Yin and Xiaoqing Wen. Dynamical properties of a fractional reaction-diffusion trimolecular biochemical model with autocatalysis. Advances in Difference Equations 2017, no. 1 (2017): 1–20.

[18] Wenting Zhang, Wei Xu, Qin Guo and Hongxia Zhang. Bifurcations in a time-delayed birhythmic biological system with fractional derivative and levy noise. International Journal of Bifurcation and Chaos 31, no. 16 (2021): 2150244.

[19] Parvaiz Ahmad Naik, Zohreh Eskandari and Hossein Eskandari Shahraki. Flip and generalized flip bifurcations of a two-dimensional discrete-time chemical model. Mathematical Modelling and Numerical Simulation with Applications 1, no. 2 (2021): 95–101.

[20] Mahdy, A. M. S. and M. Higazy. Numerical different methods for solving the nonlinear biochemical reaction model. International Journal of Applied and Computational Mathematics 5, no. 6 (2019): 1–17.

[21] Muhammad Asad Iqbal, Syed Tauseef Mohyud-Din and Bandar Bin-Mohsin. A study of nonlinear biochemical reaction model. International Journal of Biomathematics 9, no. 05 (2016): 1650071.

[22] Zafar, Z. A., Kashif Rehan, Muhammad Mushtaq and Muhammad Rafiq. Numerical modeling for nonlinear biochemical reaction networks. Iranian Journal of Mathematical Chemistry 8, no. 4 (2017): 413–423.

[23] Zhong Zhao, Li Yang and Lansun Chen. Bifurcation and chaos of biochemical reaction model with impulsive perturbations. Nonlinear Dynamics 63, no. 4 (2011): 521–535.

[24] Ali Akgul and SarbazH A. Khoshnaw. Application of fractional derivative on non-linear biochemical reaction models. International Journal of Intelligent Networks 1 (2020): 52–58.

[25] Said SEH Elnashaie, Zhongxiang Chen, Parag Garhyan, Pradeep Prasad and Andres Mahecha-Botero. Practical implications of bifurcation and chaos in chemical and biological reaction engineering. International Journal of Chemical Reactor Engineering 4, no. 1 (2006).

[26] Jean Maquet, Christophe Letellier and Luis A. Aguirre. Scalar modeling and analysis of a 3D biochemical reaction model. Journal of Theoretical Biology 228, no. 3 (2004): 421–430.

[27] Christophe Letellier. Topological analysis of chaos in a three-variable biochemical model. Acta Biotheoretica 50, no. 1 (2002): 1–13.

[28] Torben Geest, Curtis G. Steinmetz, Raima Larter and Lars F. Olsen. Period-doubling bifurcations and chaos in an enzyme reaction. The Journal of Physical Chemistry 96, no. 14 (1992): 5678–5680.

[29] Miguel A. Aon, Sonia Cortassa and David Lloyd. Chaotic dynamics and fractal space in biochemistry: Simplicity underlies complexity. Cell Biology International 24, no. 8 (2000) 581–587.

Chapter 4

Dynamical Analysis of Memristor Based KTZ Neuron Model with Fractional Difference Operator

In order to gain a deeper understanding of the intricate mechanisms within the human brain, neurons, which are the fundamental units of the nervous system, are often used as models. Neurons play a crucial role in conveying electrical information throughout our bodies. To investigate the dynamics of neurons, a discrete–time fractional order KTz map is employed in this study. The stability properties of the neuron model are assessed numerically at its equilibrium position by analyzing eigenvalues. Chaotic behavior in the neuron model is visualized through bifurcation diagrams and phase plane plots to examine the breakdown of the system. The influence of the fractional order on system behavior and the presence of coexisting attractors are also explored to enhance our comprehension of the system. To introduce memory into the map, three newly proposed discrete fractional flux controlled memristors with cubic, periodic, and exponential functions are incorporated. These flux controlled memristor elements utilize three different memductance functions. Bifurcation analysis is conducted to qualitatively analyze the three cases. The computational results presented in this study demonstrate the complex dynamic behavior exhibited by the neuronal models.

4.1 Introduction

Neuroscience encompasses a vast area of study focused on the intricate functioning of nervous systems that govern organisms' entire bodies. Over the course of many years, advancements in experimental methods have greatly accelerated progress in this field. The significance of neuroscience extends beyond comprehending the human nervous system; it plays a crucial role in developing effective medications and strategies for controlling various health conditions such as brain tumors, strokes, immune disorders, epilepsy, and more. Neuroscience comprises several sub-fields, including behavioral neuroscience (exploring how human behavior is influenced by the brain), cognitive neuroscience (investigating thought processes and control), cultural neuroscience (examining the relationship between culture, genetics, and psychological processes), cellular and molecular neuroscience (studying nerve cell functions at the cellular and molecular levels), neuroimaging (using medical imaging techniques for disease diagnosis and studying brain functions), and neuroengineering (developing engineering techniques to understand, repair, and replace neural systems). These sub-fields have been established to foster expertise in different scientific domains.

The history of neuroscience traces back to ancient times when Greek scholars attempted to comprehend the workings and role of the brain in humans. In the 19th century, a French anatomist named Paul Broca provided explanations for specific functions performed by different brain regions. Von Helmholtz illustrated the concept of electric impulses in nerve cells, and it was in the 20th century that the study of neurons and their categorization gained significant attention from biologists.

Neurons serve as the fundamental units of the nervous system, forming interconnected patterns that facilitate the transmission of information from the brain to other parts of an organism. Neuronal function is highly complex, with dense packing of over 10,000 cell connections. Despite extensive research, neurologists have yet to provide a complete and satisfactory understanding of how neurons function and interact to produce behavioral changes in organisms. While the size of neurons may vary, their essential features remain the same. A neuron consists of three crucial regions: dendrites (branch-like structures) that receive input signals from other neurons, the soma (cell body) acting as a central processing unit, and the axon that transmits processed information as output to other neurons through synapses (junctions between neurons). Though this simplified description may not accurately represent the biological intricacies of neuronal function, it provides an overview of input reception and output transmission.

The complex networks of neurons play a vital role in human decision-making processes. Neurons receive information from sensory organs such as the eyes, ears, nose, tongue, and skin, which is then sent to the brain for decision-making purposes. It is important to note that the transfer of information involves not just a finite number of neurons, but rather over 100 billion neurons in the brain that participate in every human activity. For example, when tasting something sweet, receptor neurons in the tongue send information to determine its tastiness. This decision-making information is then passed through multiple layers of neurons until a specific neuron fires, indicating the output response and influencing the perception of the sweetness. Therefore, neurology is a constantly evolving field, with new research findings continuously enhancing our knowledge and understanding of both the biological and psychological aspects of human functioning. Communication between neurons occurs through chemical synapses and electrical synapses.

4.2 Mathematical Models of Neuron

In the realm of neuroscience, understanding the intricate networks of neurons and their complex dynamics is a daunting task for humans, and conducting experiments alone is insufficient and financially impractical to capture the complete behavioral aspect. This is where mathematics and computation play a pivotal role by providing theoretical models of biological systems and enabling investigation of their intricate and fascinating dynamics. The integration of mathematical principles into neuroscience began in the mid-20th century and has since witnessed tremendous growth in interdisciplinary research. The concept of modeling neurons using dynamical systems was initially explored by Norbert Wiener in his book [6], which delved into the relationship between neuron function and computers. John von Neumann [5] also discussed brain functions while developing programmable digital computers in his book. However, it was Alan Hodgkin and Andrew Huxley[4] who transformed the concept into reality by developing a cellular-level model of neurons. Their groundbreaking work laid the foundation for subsequent neuron models and gave rise to the field of mathematical neuroscience, attracting researchers to pursue careers in this domain.

Mathematical models in neuroscience can be categorized as experimentally verifiable models or models that replicate the mechanisms of biological systems. The former type is crucial for understanding the biological aspects of systems, while the latter focuses on exploring the physiological behavior of systems using parameters that describe the underlying mechanisms. Mathematically speaking, neuron models are constructed to investigate the dynamic characteristics of neurons, such as spiking and bursting patterns. Recently, there has been a shift in interest towards the use of fractional order operators in modeling, owing to their ability to introduce a memory factor into system analysis. Fractional order systems increase the degrees of freedom by one, thereby exerting diverse impacts on system behavior. Their suitability for describing processes with memory makes them ideal for constructing neuron models to better understand their biophysical nature.

Numerous fractional order models of neurons have been proposed in the literature. For instance, the realization of fractional order neurons was performed in [7, 15], while another explored the dynamics of the Hindmarsh-Rose model of neurons using fractional derivatives [16]. The applications of fractional order derivatives were investigated in relation to the patterns of neurons from the Morris-Lecar model [18] and the bursting behavior of neurons in the FitzHugh-Rinzel model [19]. Contributions also include the synchronization of fractional order neuronal models [17] and the analysis of neuron firing patterns using the FitzHugh-Nagumo model [20]. Overall, mathematics has become an indispensable tool in neuroscience, enabling the construction of theoretical models and facilitating the exploration of complex neuronal dynamics. This interdisciplinary approach has significantly advanced our understanding of the brain and its intricate workings.

4.2.1 Discrete Fractional KTZ model

Map-based models of neurons have emerged as a result of the McCulloch-Pitts model [9] concept, offering a simpler and computationally efficient representation of neuronal dynamics. These models have garnered significant interest due to their ability to capture recurring dynamical changes and provide clearer illustrations. Among the notable map-based neuron models, the Kinouchi-Tragtenberg (KT) model [8] holds particular importance. The KT model describes neuronal functions using hyperbolic functions and was initially proposed as a two-dimensional map.

$$\begin{cases} x(n+1) = & F\left(\dfrac{x(n) - \mu_1 y(n) + \mathcal{H} + I(n)}{\mu_3}\right), \\ y(n+1) = & x(n), \end{cases} \tag{4.1}$$

where $x(n)$ and $y(n)$ denote the neuron's membrane potential and recovery variable, respectively. The function $F(x)$ is studied considering the hyperbolic tangent $\tanh\left(\frac{x}{\mu_3}\right)$, $\frac{1}{\mu_3}$ is the gain parameter, and μ_1 is a real constant, while $I(n)$ is the external input. Considering the parameter $\mathcal{H} \equiv z(n)$ in map (4.1), the extended model is obtained as

$$\begin{cases} x(n+1) = & \tanh\left(\dfrac{x(n) - \mu_1 y(n) + z(n) + I(n)}{\mu_3}\right), \\ y(n+1) = & x(n), \\ z(n+1) = & (1-\mu_4)z(n) - \mu_5(x(n) - \mu_6), \end{cases} \tag{4.2}$$

where the inverse recovery time of total ionic current $z(n)$ is represented by μ_4 and μ_5, μ_6 are the control parameters. The KTz model in (4.2) is considered for study by transforming it into a discrete fractional system given by

$$\begin{cases} \Delta^{\vartheta} x(n) = & -\tanh\left(\dfrac{x(n-1+\vartheta) - \mu_1\, y(n-1+\vartheta) + z(n-1+\vartheta) + \mu_2}{\mu_3}\right) - x(n-1+\vartheta), \\ \Delta^{\vartheta} y(n) = & x(n-1+\vartheta) - y(n-1+\vartheta), \\ \Delta^{\vartheta} z(n) = & -\mu_4 z(n-1+\vartheta) - \mu_5(x(n-1+\vartheta) - \mu_6), \end{cases} \tag{4.3}$$

where Δ^{ϑ} is the Caputo type fractional difference operator with $0 < \vartheta \leq 1$, $\mu_i > 0, i = 1,2,3,4,5,6$ are real parameters. The analyses in this chapter are performed considering a constant external input $I(n) = \mu_2$. The above proposed model of the KTz neuron model cannot be investigated in its present form, so we employ the fractional sum equation method proposed in [2], and arrive at the numerically suitable form given by

$$\begin{cases} x(n) = & x(0) + \dfrac{1}{\Gamma(\vartheta)} \sum\limits_{k=1}^{n} \dfrac{\Gamma(n-k+\vartheta)}{\Gamma(n-k+1)} \Big[-\tanh\left(\dfrac{x(k-1) - \mu_1\, y(k-1) + z(k-1) + \mu_2}{\mu_3}\right) \\ & -x(k-1) \Big], \\ y(n) = & y(0) + \dfrac{1}{\Gamma(\vartheta)} \sum\limits_{k=1}^{n} \dfrac{\Gamma(n-k+\vartheta)}{\Gamma(n-k+1)} \Big[x(k-1) - y(k-1) \Big], \\ z(n) = & z(0) + \dfrac{1}{\Gamma(\vartheta)} \sum\limits_{k=1}^{n} \dfrac{\Gamma(n-k+\vartheta)}{\Gamma(n-k+1)} \big[-\mu_4 z(k-1) - \mu_5 (x(k-1) - \mu_6) \big], \end{cases} \tag{4.4}$$

where $n = 1, 2, 3 \cdots$.

4.2.2 Stability and Instability of KTZ Neuron Model

In order to discuss the stability properties of the proposed system, we shall recall an important result in the form of the theorem proposed by [1].

Theorem 4.1 *[1]Let $\vartheta \in (0,1)$ and $\Psi \in \mathbb{R}^{k \times k}$. Let*

$$B^{\vartheta} = \left\{ \lambda \in \mathbb{C} : |\lambda| < \left(2\cos\left(\frac{|arg\lambda| - \pi}{2 - \vartheta} \right) \right)^{\vartheta} \text{ and } |arg\lambda| > \frac{\vartheta \pi}{2} \right\}.$$

If all the eigenvalues of Ψ lies in B^{ϑ}, then the system (4.3) *is asymptotically stable. The system* (4.3) *is unstable if $\lambda \in \mathbb{C} \backslash cl(B^{\vartheta})$ for an eigenvalue of Ψ.*

Asymptotic Stability of the System (4.3):
To investigate the stability of the KTZ model of neurons at a non-trivial equilibrium state, we will perform numerical analysis using specific parameter values. For the given parameters $\mu_1 = 1.25$, $\mu_2 = 0.4$, $\mu_3 = 1.25$, $\mu_4 = 0.5$, $\mu_5 = 0.75$, $\mu_6 = 0.85$, and $\vartheta = 0.85$, we will focus on the real-valued equilibrium state obtained as $E_0 = (0.5328118052, 0.5328118052, 0.4757822923)$. To assess the stability of the system, we will utilize Theorem 4.1, which relies on the analysis of eigenvalues. Firstly, we need to evaluate the variational matrix of the KTZ model (4.3) at the equilibrium state.

$$V(E_0) = \begin{bmatrix} -0.4271107358 & -0.7161115802 & 0.5728892642 \\ 1 & -1 & 0 \\ -0.75 & 0 & -0.5 \end{bmatrix}. \tag{4.5}$$

The eigenvalues corresponding the variational matrix (4.5) are $\beta_{1,2} = -0.6125867865 \pm i1.025274209$ and $\beta_3 = -0.7019371641$. Numerical calculations based on the conditions in Theorem 4.1 yield,

$$\begin{aligned} \left(2\cos\left(\frac{|arg(-0.6125867865 \pm i\, 1.025274209)| - \pi}{2 - 0.85} \right) \right)^{0.85} &= 1.204966401 \\ \left(2\cos\left(\frac{|arg(-0.7019371641)| - \pi}{2 - 0.85} \right) \right)^{0.85} &= 1.802499999. \end{aligned} \tag{4.6}$$

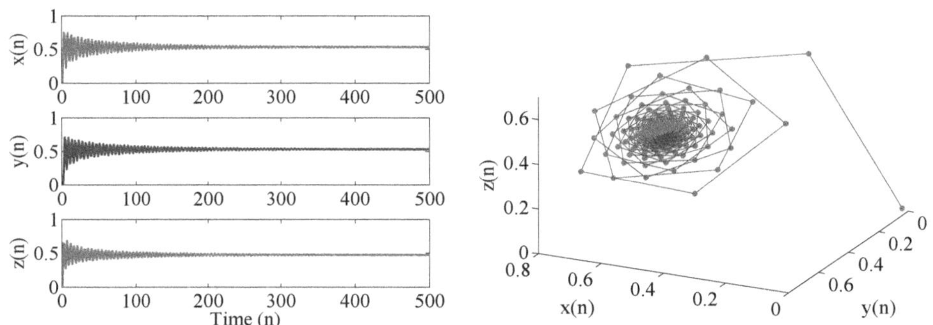

Fig. 4.1: Time series and phase plane plots of the system (4.3).

Absolute values of the eigenvalues are $|-0.6125867865 \pm i\ 1.025274209| = 1.194340728$ and $|-0.7019371641| = 0.7019371641$. Similarly, arguments of the eigenvalues are $|arg(-0.6125867865 \pm i\ 1.025274209)| = 2.109365104$ and $|arg(-0.7019371641)| = 3.141592654$. Thus, after comparison of the numerically evaluated values based on Theorem (4.1), one can observe that

$$|\beta_{1,2,3}| < \left(2\cos\left(\frac{|arg\beta_{1,2,3}| - \pi}{2 - \vartheta}\right)\right)^{\vartheta},$$

is satisfied and for the second condition it is obvious that absolute values of the arguments of $\beta_{1,2,3}$ are clearly greater than $\frac{0.85\pi}{2} = 1.335714286$. The stability conditions for the system are satisfied analytically, and numerical simulations provide further support for the results. Time series plots of the state variables and 3D phase plane plots are presented in Figure 4.1, demonstrating the convergence of the system over time. The phase plane diagram shows trajectories starting from the initial state spiraling inward towards the equilibrium state. Similarly, the time series trajectories exhibit initial oscillations before decaying towards their respective equilibrium states. These visualizations provide additional evidence for the stability of the system.

Instability of the System (4.3):
Assuming the parameters of the system (4.3) take the following values: $\mu_1 = 0.65$, $\mu_2 = 1.8$, $\mu_3 = 1.05$, $\mu_4 = 0.45$, $\mu_5 = 1.75$, $\mu_6 = 0.25$, and $\vartheta = 0.45$, with an initial state of $(x(0), y(0), z(0)) = (0.01, 0.02, 0.02)$, we can determine the real-valued equilibrium states as $E_0 = (0.5846962090, 0.5846962090, -1.301596368)$. To verify the stability of the system, we will utilize Theorem 4.1 by evaluating the variational matrix of the system (4.3) at the equilibrium state.

$$V(E_0) = \begin{bmatrix} -0.3732091967 & -0.4074140221 & 0.6267908033 \\ 1 & -1 & 0 \\ -1.75 & 0 & -0.45 \end{bmatrix}. \tag{4.7}$$

The eigenvalues corresponding to the variational matrix (4.7) are $\beta_{1,2} = -0.4770421800 \pm i\ 1.199443041$ and $\beta_3 = -0.8691248368$. Numerical calculations based on the conditions in Theorem 4.1 yield,

$$\begin{aligned}\left(2\cos\left(\frac{|arg(-0.4770421800 \pm i\ 1.199443041)| - \pi}{2-0.45}\right)\right)^{0.45} &= 1.013715355\\ \left(2\cos\left(\frac{|arg(-0.86912483681)| - \pi}{2-0.85}\right)\right)^{0.85} &= 1.802499999.\end{aligned} \tag{4.8}$$

Absolute values of the eigenvalues are $|-0.4770421800 \pm i\ 1.199443041| = 1.290826421$ and $|-0.8691248368| = 0.8691248368$. Similarly, arguments of the eigenvalues are $|arg(-0.4770421800 \pm i\ 1.199443041)| = 1.949335422$ and $|arg(-0.8691248368)| = 3.141592654$. Thus, after comparison of the numerically evaluated values based on Theorem (4.1), one can observe that from 4.8

$$|\beta_3| < \left(2\cos\left(\frac{|arg\beta_3| - \pi}{2-\vartheta}\right)\right)^{\vartheta}$$

for β_3 and

$$|\beta_{1,2}| > \left(2\cos\left(\frac{|arg\beta_3| - \pi}{2-\vartheta}\right)\right)^{\vartheta}$$

for $\beta_{1,2}$ and for the second condition it is obvious that absolute value of the arguments of $\beta_{1,2,3}$ are clearly greater than $\frac{0.85\pi}{2} = 0.7071428571$. The failure of the stability condition from Theorem 4.1 analytically indicates instability in the system. To numerically verify the instability, time series and phase plots for the given parameter values are presented in Figure 4.2. These visualizations provide evidence of the unstable behavior of the system.

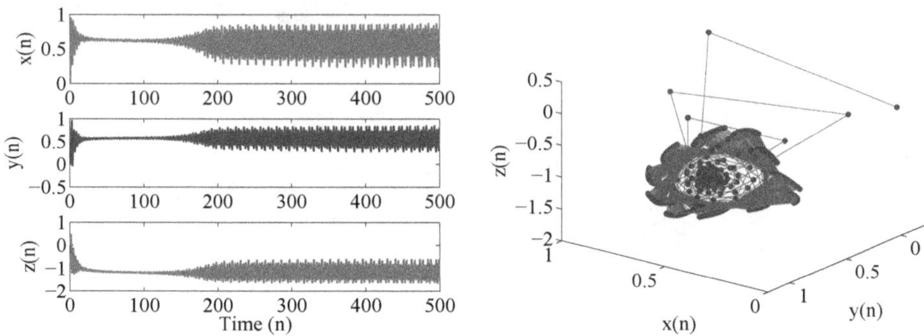

Fig. 4.2: Time series and phase plane plots of the system (4.3).

Figure 4.2 depicts the unstable behavior of the system for the given parameter values. The time series trajectories in the figure exhibit an initial decay, followed by an oscillatory behavior that persists over increasing time. The corresponding phase plane plots demonstrate trajectories spiraling outward from the equilibrium state, further illustrating the unstable nature of the system.

4.2.3 Chaotic Dynamics of Discrete Fractional KTZ Neuron Model

This section aims to investigate the chaotic behavior of the neuron model by varying the parameter μ_3. The model considers the fractional difference KTZ neuron model and employs bifurcation diagrams and Lyapunov exponents to illustrate the chaotic dynamics. The simulations are performed by varying μ_3 in the range of $[0, 1.5]$ while keeping other parameters fixed. The initial state is set to $(x(0), y(0), z(0)) = (1, 2, 2)$. Figures 4.3 showcase the system's transitions through 3-D phase plane plots, providing visual insights into the changes occurring with different values of μ_3. The bifurcation diagram in Figure 4.3 illustrates the system's behavior as μ_3 is varied. It shows a reverse bifurcation, where the system exhibits chaotic behavior for lower values of μ_3 and transitions to a more predictable and stable behavior for higher values of μ_3.

To further analyze the transitional phases of the system, the maximum Lyapunov exponents are plotted in Figure 4.3. Trajectories above the zero line indicate chaotic behavior, while trajectories below the zero line represent periodic behavior. To gain a better understanding of the system's state changes, 3-D phase plane plots are presented at specific values of μ_3: 0.325, 0.75, 0.95, 1.15, and 1.4. Figure 4.4 demonstrates the transition from chaotic behavior to periodic behavior. The first plot shows chaotic dynamics, while the second plot exhibits closed orbits with spirals moving outwards. Figure 4.5 focuses on the transformation from unstable closed orbits to a stable trajectory spiraling inwards towards an equilibrium point.

These visualizations help to understand the dynamic behavior of the neuron model as μ_3 is varied, highlighting the presence of chaos, periodicity, and stability within the system.

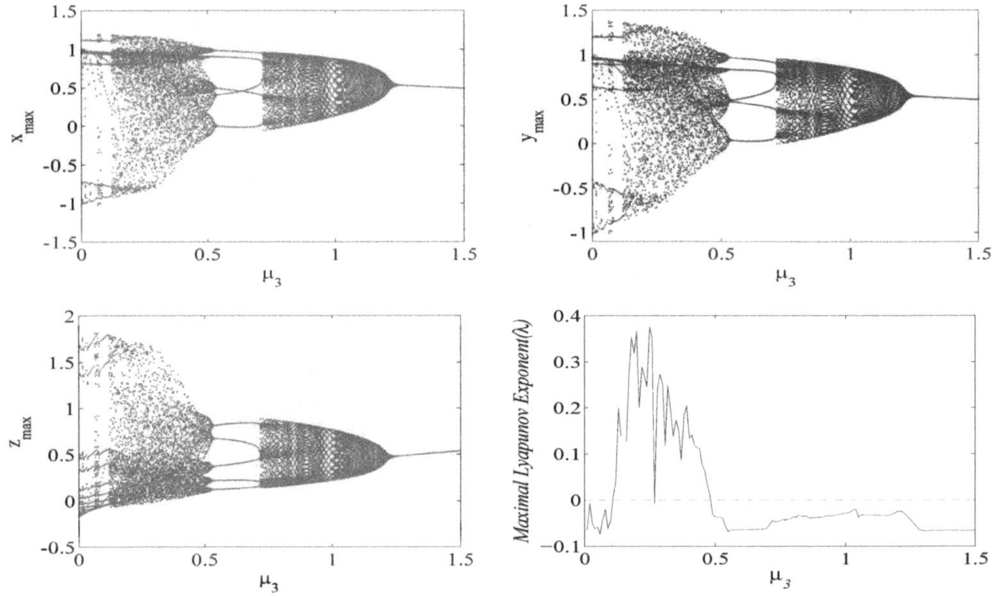

Fig. 4.3: Bifurcation diagrams and maximum Lyapunov exponent of the system (4.3).

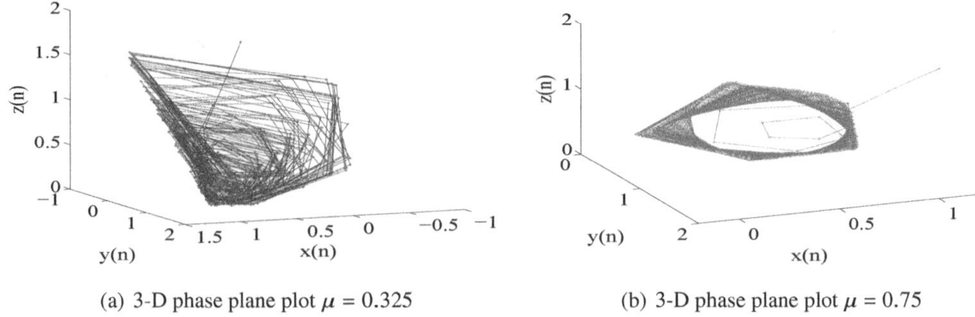

(a) 3-D phase plane plot $\mu = 0.325$

(b) 3-D phase plane plot $\mu = 0.75$

Fig. 4.4: 3-D phase plane plots of system (4.3).

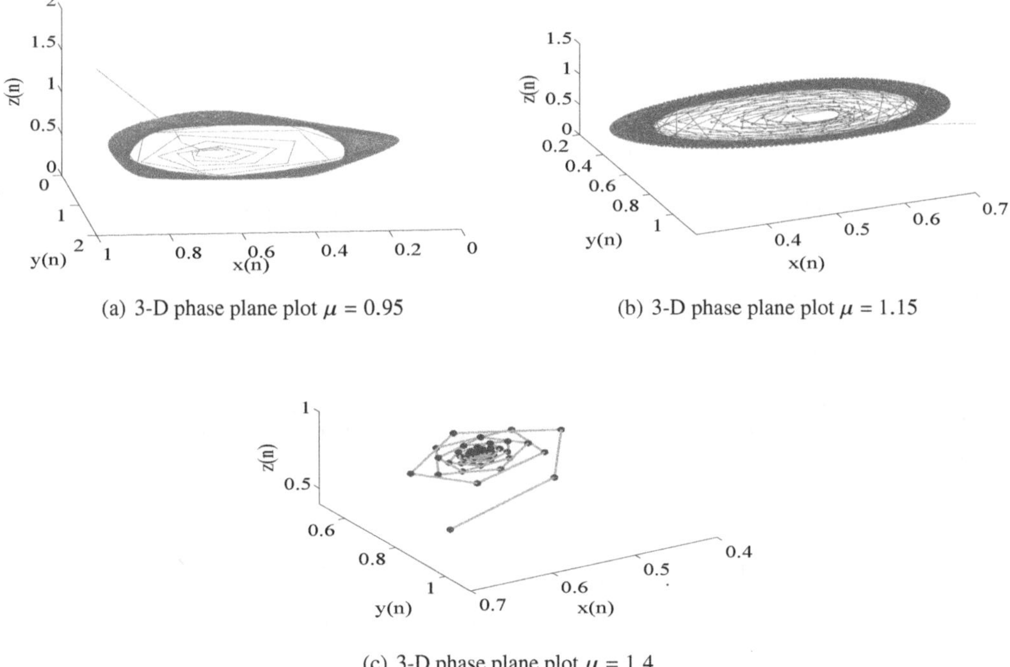

(a) 3-D phase plane plot $\mu = 0.95$

(b) 3-D phase plane plot $\mu = 1.15$

(c) 3-D phase plane plot $\mu = 1.4$

Fig. 4.5: 3-D phase plane plots of system (4.3).

4.2.4 Impact of Fractional Order ϑ on the System (4.3)

This section investigates the influence of varying fractional orders on the dynamic behavior of the KTZ neuron model (4.3). The analysis focuses on bifurcation diagrams obtained for a fixed set of parameters: $\mu_1 = 0.65, \mu_2 = 1.8, \mu_3 = 1.05, \mu_4 = 0.35, \mu_5 = 1.75, \mu_6 = 0.25$. The initial state is set to $(x(0), y(0), z(0)) = (0.01, 0.02, 0.02)$, and the fractional order ϑ is varied within the range $(0, 1)$.

Since the model incorporates fractional difference operators, it is important to explore the impact of different fractional orders on the system's dynamics. The bifurcation diagrams in Figure 4.6 illustrate the transitions of the system from stability to chaos as the fractional order ϑ is varied. The presence of

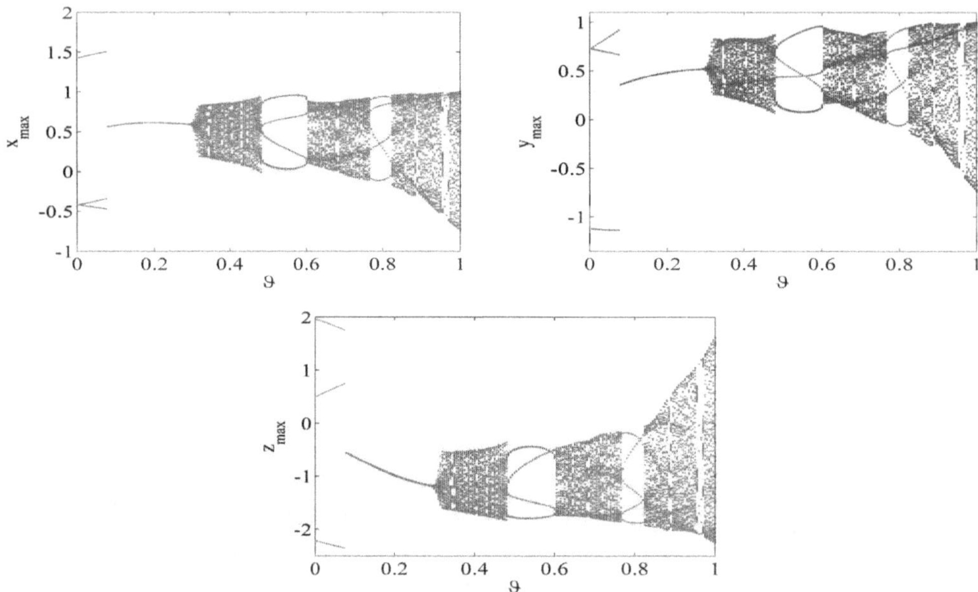

Fig. 4.6: Bifurcation diagrams of the system (4.3) for varying fractional order (ϑ).

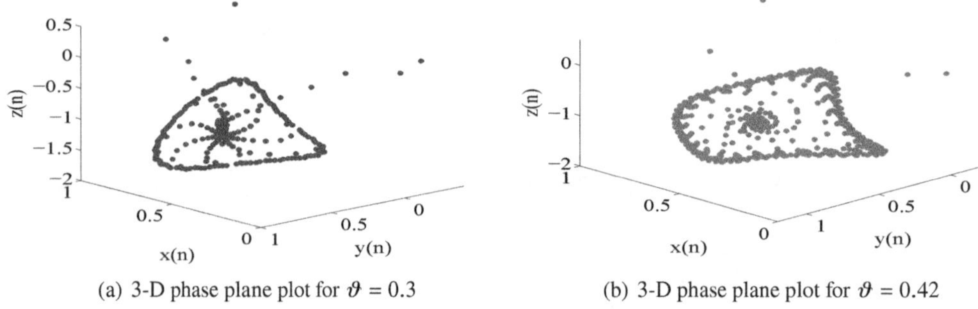

(a) 3-D phase plane plot for $\vartheta = 0.3$

(b) 3-D phase plane plot for $\vartheta = 0.42$

Fig. 4.7: 3-D phase plane plots of system (4.3).

periodic windows within the bifurcation diagram visually depicts the system's transition. These windows indicate regions where the system exhibits periodic behavior amidst the chaotic dynamics. By analyzing the bifurcation diagrams for different fractional orders, one can gain insights into the influence of the fractional order parameter on the behavior of the KTZ neuron model.

To further analyze the impact of fractional orders on the dynamical behavior of the KTZ neuron model, 3-D phase plane trajectories are examined at different fractional order values ϑ. These trajectories provide a visual representation of how the system evolves under varying fractional orders. Figure 4.7 presents phase plane trajectories that illustrate the transformation of the system from a state of stable behavior to the formation of unstable closed orbits. This transition is evident as the fractional order ϑ increases. The trajectories depict the system's progression from regular oscillations to the emergence of chaotic dynamics.

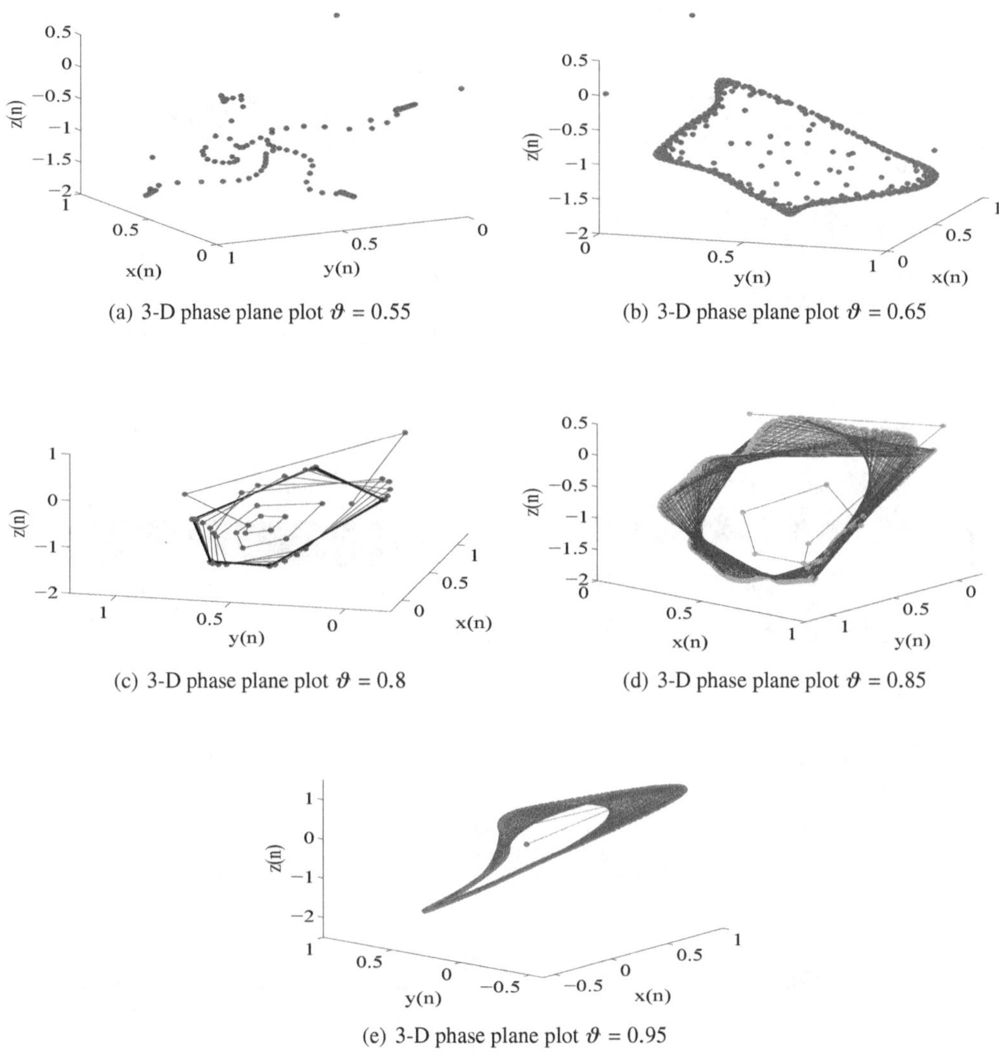

(a) 3-D phase plane plot $\vartheta = 0.55$

(b) 3-D phase plane plot $\vartheta = 0.65$

(c) 3-D phase plane plot $\vartheta = 0.8$

(d) 3-D phase plane plot $\vartheta = 0.85$

(e) 3-D phase plane plot $\vartheta = 0.95$

Fig. 4.8: 3-D phase plane plots of system (4.3).

In Figure 4.8, the behavior of the system within a periodic window is depicted. Specifically, for a fractional order value of $\vartheta = 0.55$, the system exhibits a stagnant stage characterized by uniform oscillations. This state represents a period within the bifurcation diagram where the system's behavior becomes periodic amidst the overall chaotic dynamics. Furthermore, Figure 4.8 also demonstrates the system's behavior for higher fractional order values within the range of $\vartheta \in [0.8, 1)$. In this regime, the system transitions from an initial stage of stable behavior to the formation of unstable closed orbits.

By examining the 3-D phase plane trajectories at different fractional orders, a clearer understanding of the system's dynamics and the influence of the fractional order parameter can be obtained.

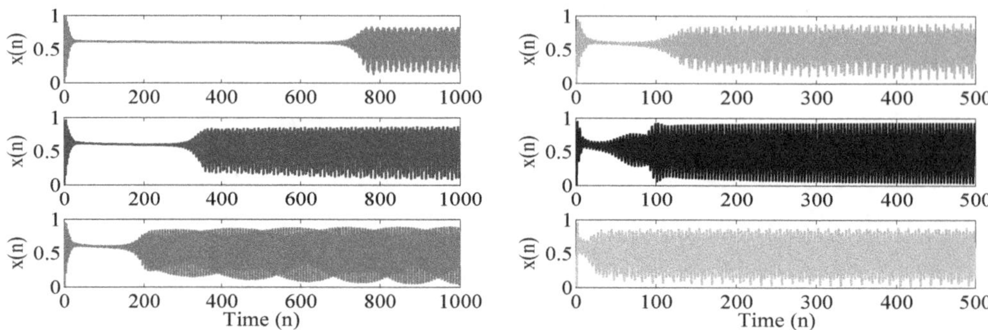

Fig. 4.9: Time series plots of state variable $x(n)$ system (4.3) for different fractional orders $\vartheta = 0.25, 0.3, 0.35, 0.4, 0.5, 0.65$.

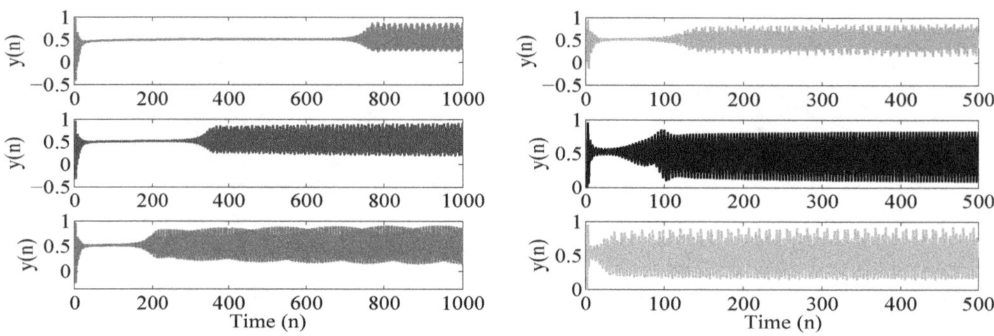

Fig. 4.10: Time series plots of state variable $y(n)$ system (4.3) for different fractional orders $\vartheta = 0.25, 0.3, 0.35, 0.4, 0.5, 0.65$.

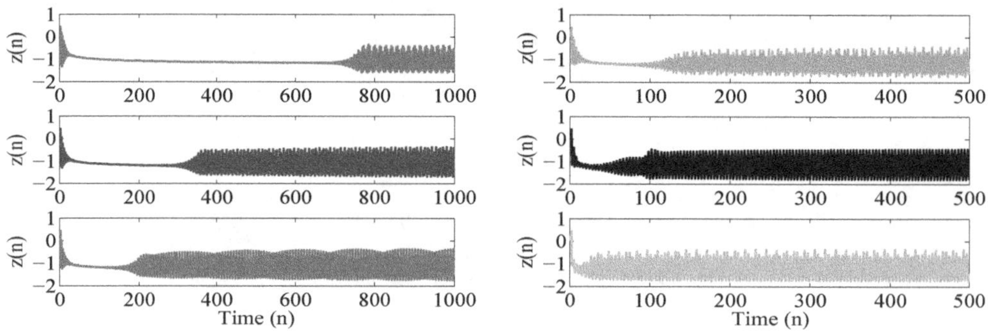

Fig. 4.11: Time series plots of state variable $z(n)$ system (4.3) for different fractional orders $\vartheta = 0.25, 0.3, 0.35, 0.4, 0.5, 0.65$.

To gain a deeper understanding of the behavior of each state variable representing the membrane potential, recovery, and iconic state of the system, individual analysis of their behavior over time is crucial. Time series trajectories are simulated for different fractional orders of the system, and the results are depicted in Figures 4.9, 4.10, and 4.11 for the state variables $x(n)$, $y(n)$, and $z(n)$, respectively. By examining the time series trajectories, certain observations can be made when the system is modeled with small fractional orders, such as $\vartheta = 0.25, 0.3$, and 0.35 as shown in the first image of Figure 4.9. Initially, the system reaches a stable state for a certain period of time. However, it eventually experiences a sudden spike, leading to oscillations. This behavior indicates that as the fractional order increases, the system tends to exhibit chaotic behavior. Therefore, the construction of models with fractional order elements provides valuable insights into the intriguing dynamics exhibited by the system.

4.2.5 Coexisting Bifurcation and Attractors

An intriguing analysis of chaotic dynamical systems involves the exploration of coexisting attractors that emerge for different initial states of the system. This analysis provides valuable insights into the coexistant nature of the system when starting from various states. In this study, the coexistence of attractors is investigated for two different initial states: $(x(0), y(0), z(0)) = (1, 2, 2)$ and $(x(0), y(0), z(0)) = (-1, -2, -2)$. The system parameters are set as $\mu_1 = 1.25$, $\mu_2 = 0.4$, $\mu_4 = 0.5$, $\mu_5 = 0.75$, $\mu_6 = 0.85$, $\vartheta = 0.85$, while μ_3 varies within the range $[0, 1.5]$. Figure 4.12 illustrates the bifurcation of the system's state variables $x(n)$, $y(n)$, and $z(n)$ for the given parameter values. It showcases the behavior of the system as μ_3 varies. Furthermore, Figure 4.13 provides a visualization of the coexisting attractors at three different values of μ_3. These findings contribute to a deeper understanding of the complex dynamics exhibited by the system when multiple attractors coexist.

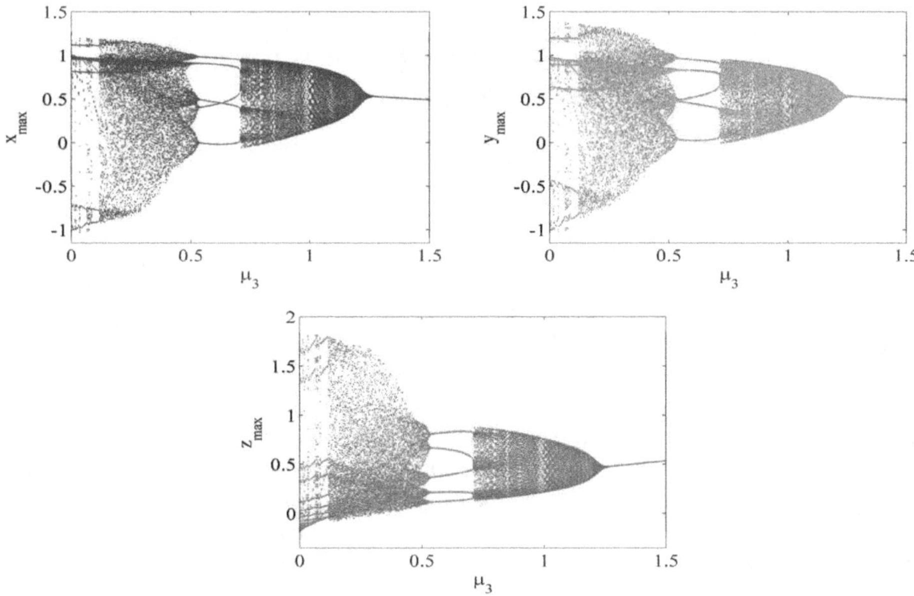

Fig. 4.12: Bifurcation diagrams of the system (4.3).

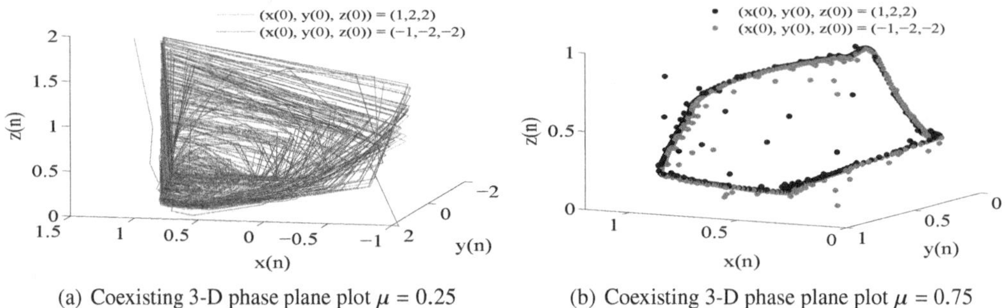

(a) Coexisting 3-D phase plane plot μ = 0.25 (b) Coexisting 3-D phase plane plot μ = 0.75

Fig. 4.13: Coexisting 3-D phase plane plots of system (4.3).

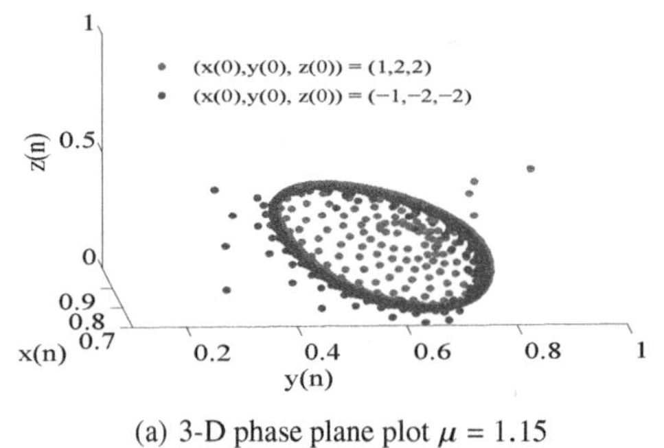

(a) 3-D phase plane plot μ = 1.15

Fig. 4.14: Coexisting 3-D phase plane plots of system (4.3).

4.3 Memristor Based KTZ Neuron Model

Memristor elements, which are characterized by the relationship between charge q and flux ϕ, are widely recognized as circuit elements. They exhibit a non-linear relationship, and can be either flux controlled or charge controlled, depending on whether their behavior is described by a function of flux or charge, respectively. Memristors are particularly notable for their memory-like properties, as their memristance and memductance are influenced by the historical values of current and voltage, respectively. While memristor elements garnered significant attention following their fabrication in HP labs in 2008 [21], they are not yet commercially available due to their high cost and fabrication challenges. In this chapter, our focus lies on flux-controlled memristive elements. We will introduce and define several flux-controlled memristive elements to enhance our understanding of their characteristics.

M1 The flux controlled memristor with cubic nonlinearity [10]: $q - \phi$ function is given by

$$q(\phi) = \eta_1\phi + \eta_2\phi^3.$$

Then,

$$W(\phi) = \eta_1 + 3\eta_2\phi^2$$

is the memductance.

M2 Exponential flux controlled memristor [11]:

$$q(\phi) = \eta_1 \left(\eta_2^{\eta_3 \phi} - 1\right)$$

The corresponding memductance is

$$W(\phi) = \eta_2^{\eta_3 \phi} \eta_1 \eta_3 \log \eta_2.$$

M3 Sine memductance [12]:

$$\begin{aligned} W(\phi) &= 1 - 0.5 \sin(\phi) \\ \dot{\phi} &= v. \end{aligned}$$

Neuron models have the ability to represent the electrical activities that arise from various external factors such as radiation, electromagnetic induction, and more. The flow of electrical and magnetic fields can be observed due to changes in the membrane potential and the exchange of ions between cells, both internally and externally. Consequently, the normal functioning of biological neurons is influenced by the presence of electrical fields. Therefore, it is crucial to investigate the dynamic characteristics of neurons under the influence of electrical fields. The effects of electromagnetic induction can be studied by incorporating memristor-based elements. Several researchers have explored the inclusion of memristive elements in neuron models, as discussed in studies such as [13] and [14]. Furthermore, circuit-based implementations of neuron models with memristors have been presented in works like [22] and [23]. In this study, we investigate the chaotic dynamics exhibited by a neuronal model that incorporates three flux-controlled memristor models with cubic nonlinearity (M1), exponential nonlinearity (M2), and a modified sine memductance (M3) that incorporates a hyperbolic cosine memductance term.

$$\begin{aligned} W(\phi) &= \eta_1 \cosh(\eta_2 \phi) \\ \dot{\phi} &= v. \end{aligned}$$

Both fractional order systems and memristors share a common characteristic: their dependence on the entire history or memory. Therefore, it is natural to consider fractional order versions of memristive elements. Several research contributions have focused on the realization and study of fractional memristor elements. For example, the authors in [24] and [25] discuss the fractional order modeling of memristors. The Hindmarsh-Rose neuron model with fractional order memristive elements is considered, and wave propagation is investigated in [26]. The hidden dynamics of the model are studied in [27]. These studies contribute to the understanding and exploration of fractional memristor elements and their applications in neuronal modeling.

4.3.1 Cubic Flux-Controlled Memristor

This subsection discusses the incorporation of fractional difference flux controlled memristive elements with cubic nonlinearity into the KTz neuronal model. Let the discrete-time fractional order flux controlled memristor model with cubic nonlinearity be

$$\begin{cases} i(t) = W_\vartheta(\Phi) \\ \Delta^\vartheta \Phi(n) = x(n - 1 + \vartheta) \\ W_\vartheta(\Phi) = \mu_8 + \mu_9 (\Phi)^2. \end{cases} \tag{4.9}$$

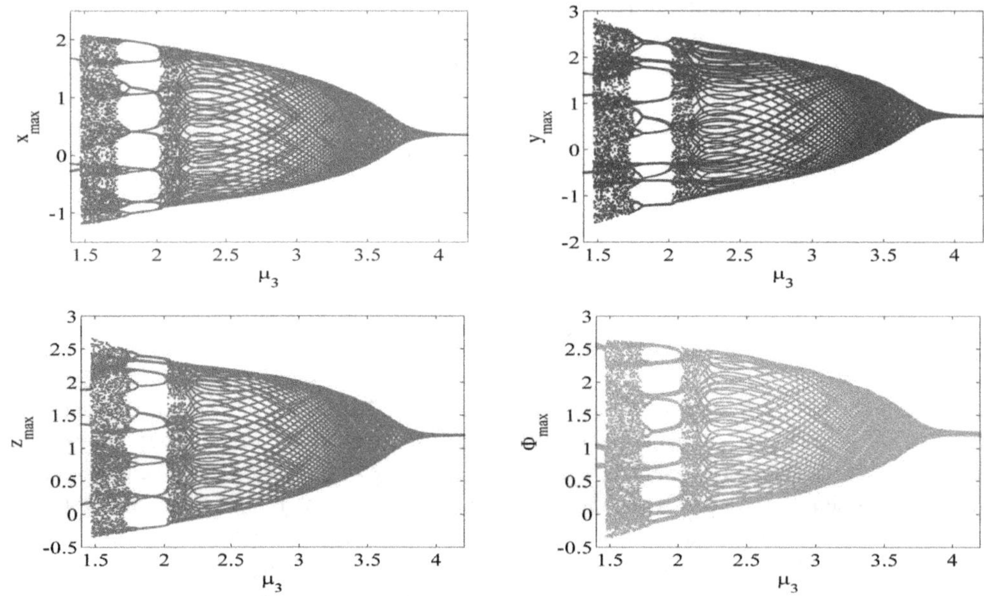

Fig. 4.15: Bifurcation diagrams of the system (4.3).

Fractional difference KTZ neuron model with nonlinear quadratic memductance function is given by

$$\begin{cases} \Delta^{\vartheta}x(n) = & -\tanh\left(\dfrac{x(n-1+\vartheta)-\mu_1\, y(n-1+\vartheta)+z(n-1+\vartheta)+\mu_2}{\mu_3}\right) \\ & -\mu_7(\mu_8+\mu_9(\Phi(n-1+\vartheta))^2)x(n-1+\vartheta)-x(n-1+\vartheta), \\ \Delta^{\vartheta}y(n) = & x(n-1+\vartheta)-y(n-1+\vartheta), \\ \Delta^{\vartheta}z(n) = & -\mu_4 z(n-1+\vartheta)-\mu_5(x(n-1+\vartheta)-\mu_6), \\ \Delta^{\vartheta}\Phi(n) = & x(n-1+\vartheta), n\in\mathbb{N}_{1-\vartheta}. \end{cases} \tag{4.10}$$

The dynamical behavior of the system (4.10) is illustrated with bifurcation diagrams simulated by varying $\mu_3 \in [1,5]$ and other parameters assuming the following values $\mu_1 = 1.85$, $\mu_2 = 0.54$, $\mu_4 = 0.5$, $\mu_5 = 0.75$, $\mu_6 = 0.85$, $\mu_7 = 0.0145$, $\mu_8 = 0.3$, $\mu_9 = 0.01$, $\vartheta = 0.2$ and initial states $(x(0), y(0), z(0), w(0)) = (1,2,2,0.01)$. Simulations are presented in Figure 4.15. The system exhibits unpredictable behavior for certain values of μ_3 later forming periodic windows exhibiting the transition to periodic states and finally stability.

The influence of the memristive elements represented with bifurcation diagrams is broken down into 3-D phase plane trajectories at three different values of μ_3 to depict the transitional states. Starting from the chaotic attractors in Figure 4.16 at $\mu_3 = 1.5$ the system then moves to the transitional phase, forming unstable closed orbits. Figure 4.17 describes the dynamics of the system after the periodic window at $\mu_3 = 2.15, 2.75, 3.15$ and the spiral inwards ensuring stability of the system.

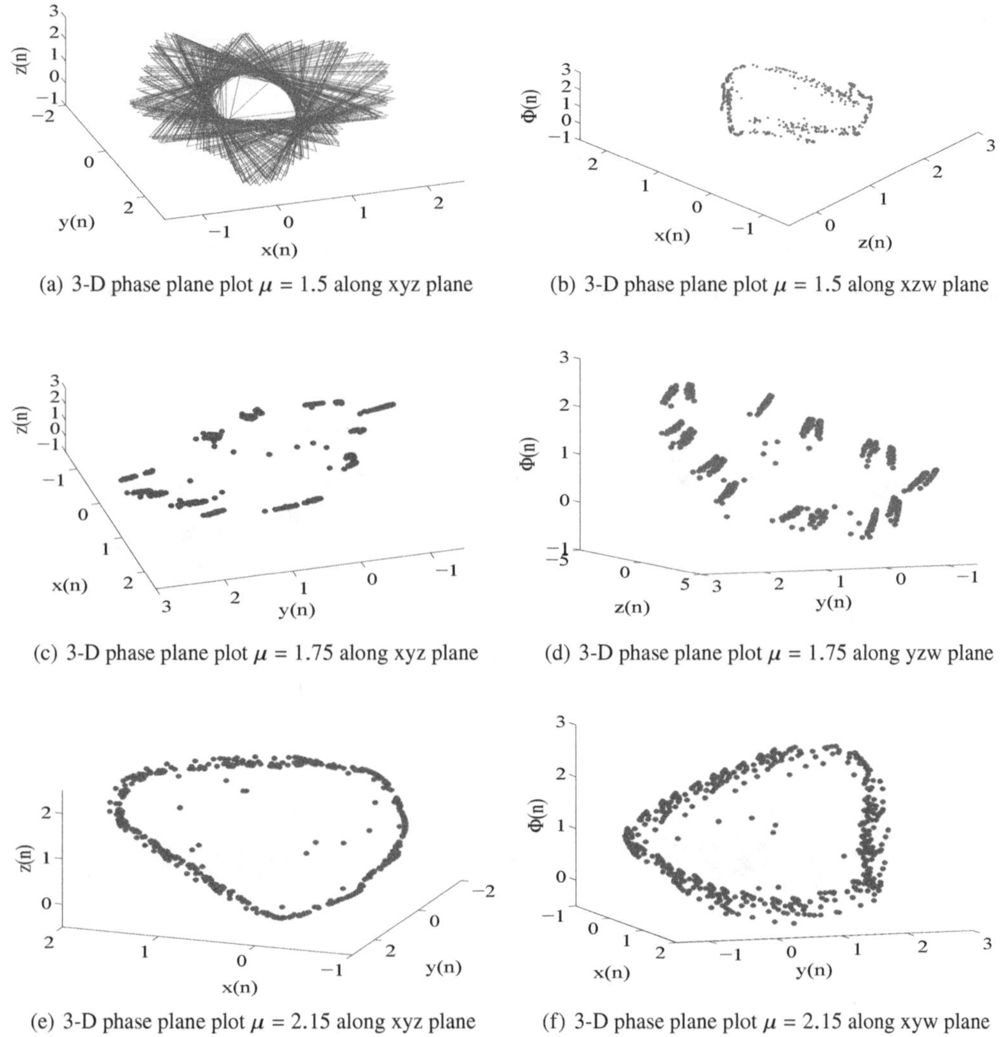

(a) 3-D phase plane plot $\mu = 1.5$ along xyz plane

(b) 3-D phase plane plot $\mu = 1.5$ along xzw plane

(c) 3-D phase plane plot $\mu = 1.75$ along xyz plane

(d) 3-D phase plane plot $\mu = 1.75$ along yzw plane

(e) 3-D phase plane plot $\mu = 2.15$ along xyz plane

(f) 3-D phase plane plot $\mu = 2.15$ along xyw plane

Fig. 4.16: 3-D phase plane plots of system (4.3) corresponding to Figure 4.15.

4.3.2 Periodic Flux-Controlled Memristor

This section aims at investigating the chaotic dynamics of the system with a periodically varying hyperbolic cosine memristor element defined by

$$\begin{cases} i(t) = W_{\vartheta}(\Phi) \\ \Delta^{\vartheta}\Phi(n) = x(n-1+\vartheta) \\ W_{\vartheta}(\Phi) = \mu_7 \cosh(\mu_8 \Phi). \end{cases} \tag{4.11}$$

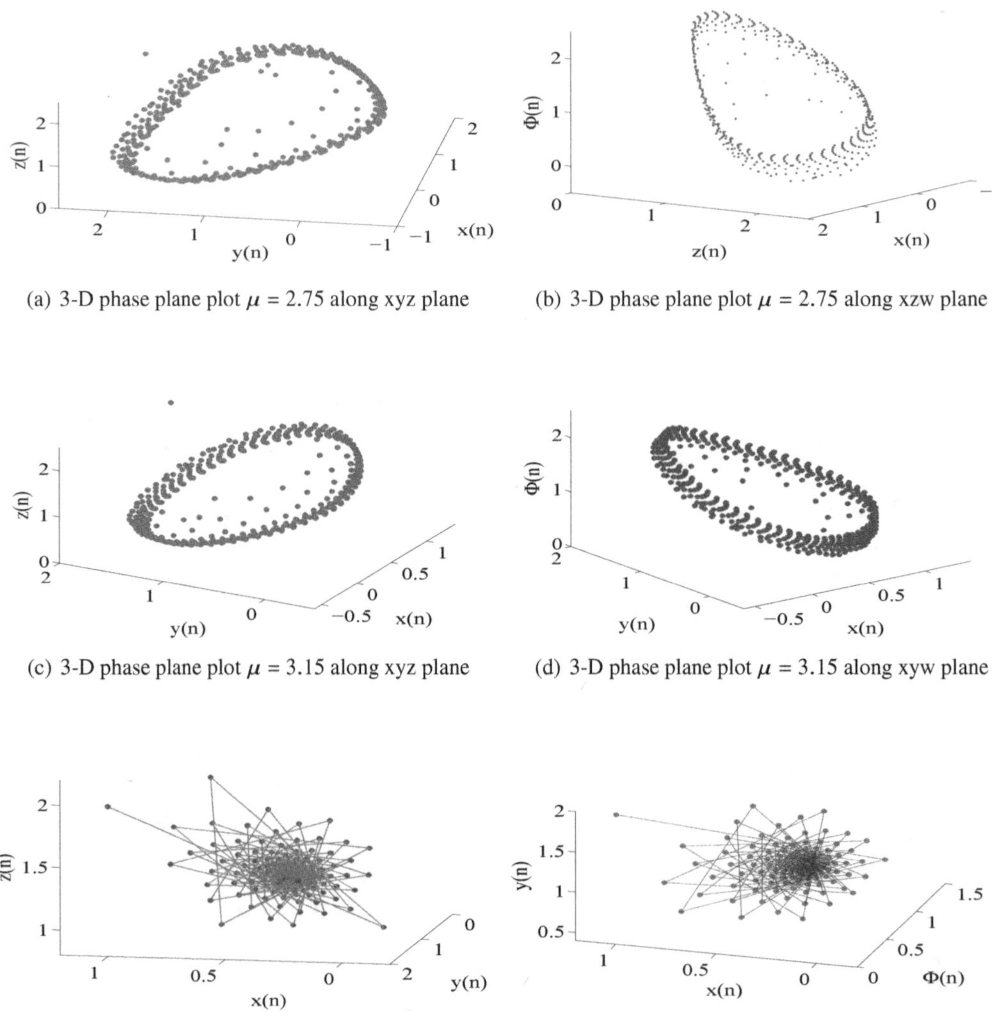

(a) 3-D phase plane plot μ = 2.75 along xyz plane

(b) 3-D phase plane plot μ = 2.75 along xzw plane

(c) 3-D phase plane plot μ = 3.15 along xyz plane

(d) 3-D phase plane plot μ = 3.15 along xyw plane

(e) 3-D phase plane plot μ = 4.1 along xyz plane

(f) 3-D phase plane plot μ = 4.1 along xyw plane

Fig. 4.17: 3-D phase plane plots of system (4.3) corresponding to Figure 4.15.

The Fractional difference KTZ neuron model with a periodic memductance function is given by

$$\begin{cases} \Delta^{\vartheta} x(n) = & -\tanh\left(\dfrac{x(n-1+\vartheta) - \mu_1\, y(n-1+\vartheta) + z(n-1+\vartheta) + \mu_2}{\mu_3}\right) \\ & -\mu_7 \cosh(\mu_8 \Phi(n-1+\vartheta)) x(n-1+\vartheta) - x(n-1+\vartheta), \\ \Delta^{\vartheta} y(n) = & x(n-1+\vartheta) - y(n-1+\vartheta), \\ \Delta^{\vartheta} z(n) = & -\mu_4 z(n-1+\vartheta) - \mu_5 (x(n-1+\vartheta) - \mu_6), \\ \Delta^{\vartheta} \Phi(n) = & x(n-1+\vartheta), n \in \mathbb{N}_{1-\vartheta}. \end{cases} \tag{4.12}$$

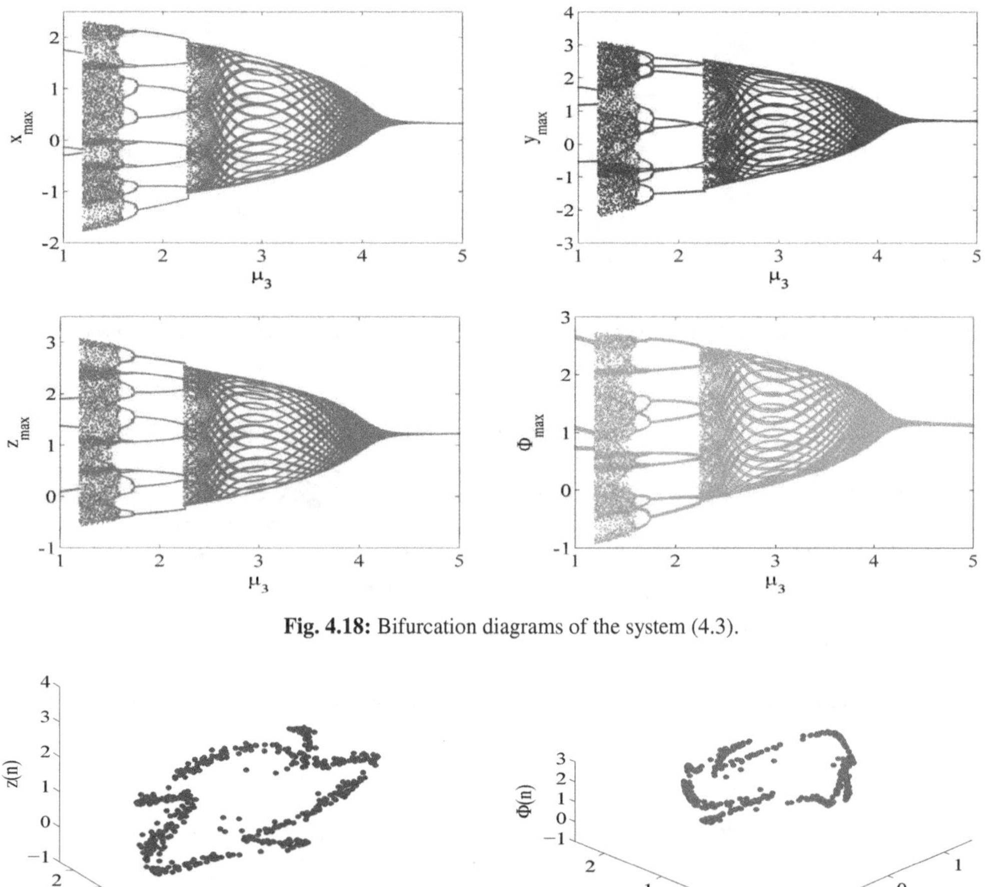

Fig. 4.18: Bifurcation diagrams of the system (4.3).

(a) 3-D phase plane plot $\mu = 1.3$ along xyz plane

(b) 3-D phase plane plot $\mu = 1.3$ along xzw plane

Fig. 4.19: 3-D phase plane plots of system (4.3) corresponding to Figure 4.18.

The dynamical behavior of the system (4.12) is illustrated with bifurcation diagrams simulated by varying $\mu_3 \in [1, 5]$ and other parameters assuming the values $\mu_1 = 1.85$, $\mu_2 = 0.54$, $\mu_4 = 0.5$, $\mu_5 = 0.75$, $\mu_6 = 0.85$, $\mu_7 = 0.05$, $\mu_8 = 0.01$, $\vartheta = 0.2$ and initial states $(x(0), y(0), z(0), w(0)) = (1, 2, 2, 0.01)$. Simulations are presented in Figure 4.18. In comparison with the system (4.10), the chaotic response of the system (4.12) is enhanced as the system exhibits chaotic behavior for values much lesser than the values of the system (4.10). This explains the chaos boosting behavior of the trigonometric and hyperbolic trigonometric functions. 3-D chaotic attractors obtained for $\mu_3 = 1.3$ are presented in Figure 4.19. The state of formation of closed orbits at $\mu_3 = 1.75, 2.3, 2.8$ is presented in Figure 4.20 and the system transition to stability is exhibited by Figure 4.21.

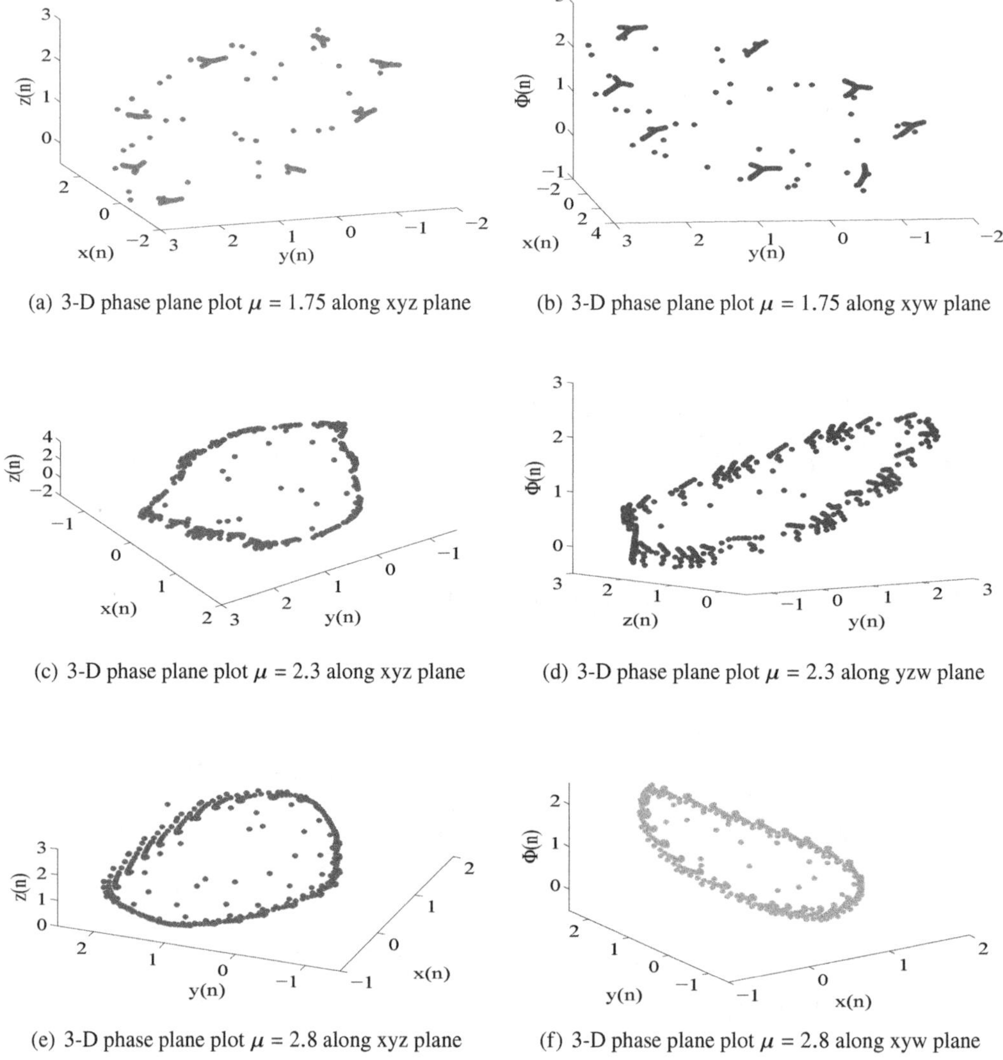

(a) 3-D phase plane plot μ = 1.75 along xyz plane

(b) 3-D phase plane plot μ = 1.75 along xyw plane

(c) 3-D phase plane plot μ = 2.3 along xyz plane

(d) 3-D phase plane plot μ = 2.3 along yzw plane

(e) 3-D phase plane plot μ = 2.8 along xyz plane

(f) 3-D phase plane plot μ = 2.8 along xyw plane

Fig. 4.20: 3-D phase plane plots of system (4.3) corresponding to Figure 4.18.

4.3.3 Exponential Flux-Controlled Memristor

An exponential flux controlled memristor element of the form

$$\begin{cases} i(t) = W_\vartheta(\Phi) \\ \Delta^\vartheta \Phi(n) = x(n-1+\vartheta) \\ W_\vartheta(\Phi) = \mu_7\mu_9 \log(\mu_8)\mu_8^{\mu_9}, \end{cases} \tag{4.13}$$

is introduced to the KTz neuronal model for analysis.

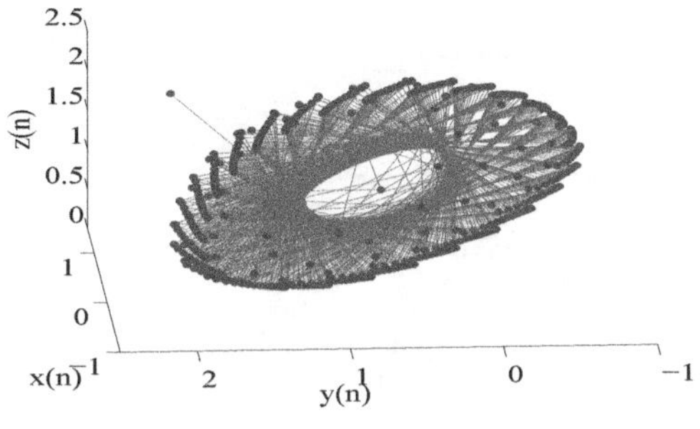

(a) 3-D phase plane plot $\mu = 3.4$ along xyz plane

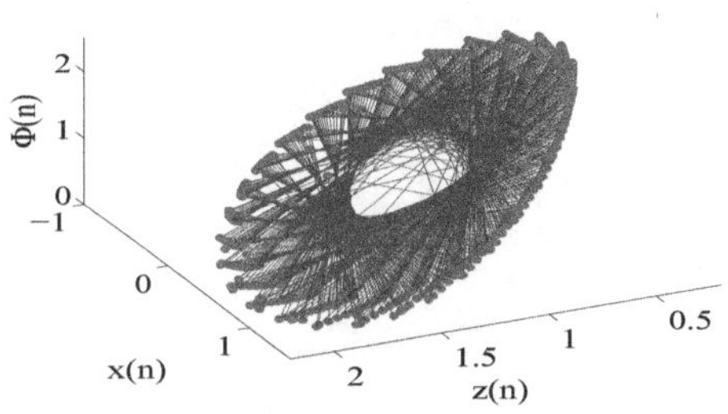

(b) 3-D phase plane plot $\mu = 3.4$ along xzw plane

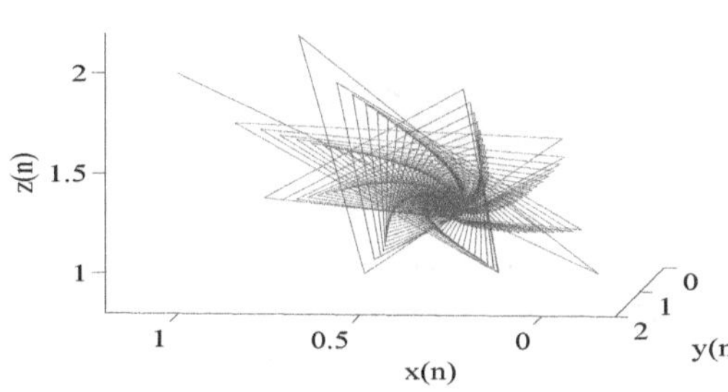

(c) 3-D phase plane plot $\mu = 4.3$ along xyz plane

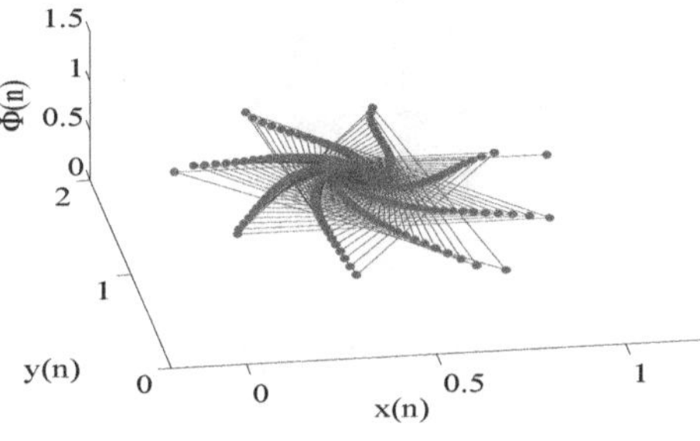

(d) 3-D phase plane plot $\mu = 4.3$ along xyw plane

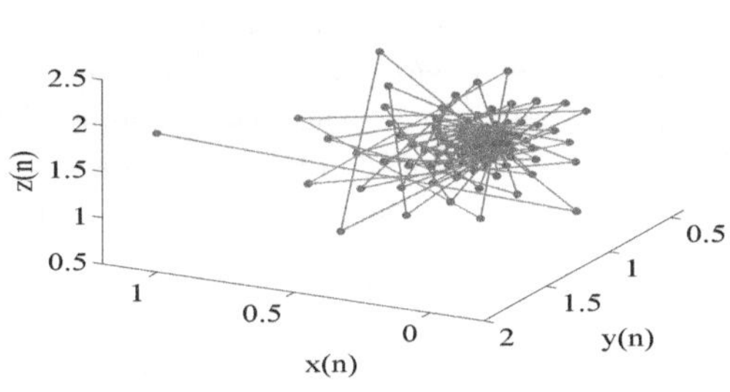

(e) 3-D phase plane plot $\mu = 4.8$ along xyz plane

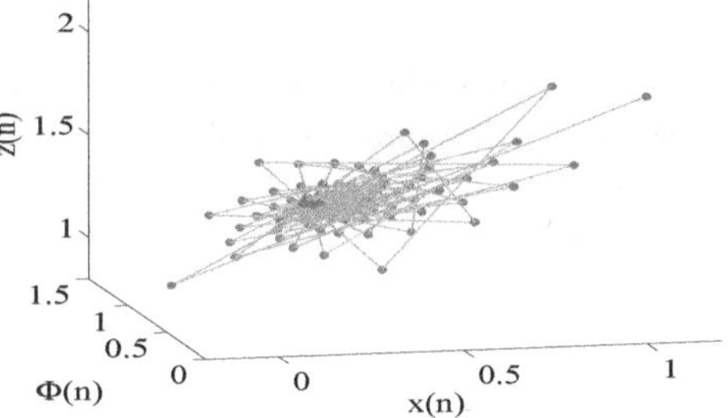

(f) 3-D phase plane plot $\mu = 4.8$ along xwz plane

Fig. 4.21: 3-D phase plane plots of system (4.3) corresponding to Figure 4.18.

The fractional difference KTZ neuron model with logarithmic memductance function is obtained by

$$\begin{cases} \Delta^{\vartheta}x(n) = & -\tanh\left(\dfrac{x(n-1+\vartheta)-\mu_1\, y(n-1+\vartheta)+z(n-1+\vartheta)+\mu_2}{\mu_3}\right) \\ & -\mu_7\mu_9\log(\mu_8)\mu_8^{\mu_9\Phi(n-1+\vartheta)}x(n-1+\vartheta)-x(n-1+\vartheta), \\ \Delta^{\vartheta}y(n) = & x(n-1+\vartheta)-y(n-1+\vartheta), \\ \Delta^{\vartheta}z(n) = & -\mu_4 z(n-1+\vartheta)-\mu_5(x(n-1+\vartheta)-\mu_6), \\ \Delta^{\vartheta}\Phi(n) = & x(n-1+\vartheta), n\in\mathbb{N}_{1-\vartheta}. \end{cases} \tag{4.14}$$

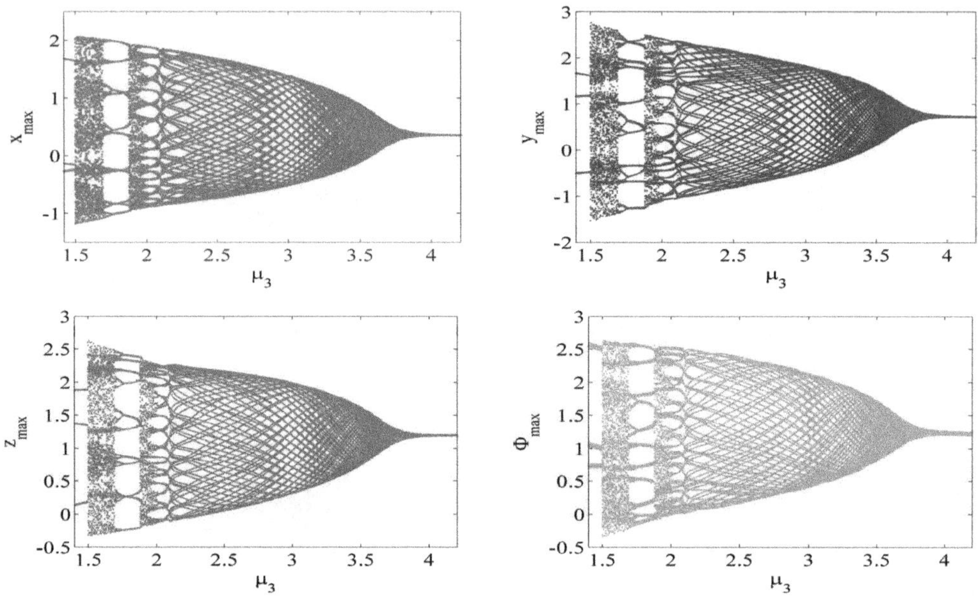

Fig. 4.22: Bifurcation diagrams of the system (4.3).

The dynamical behavior of the system (4.12) is investigated by performing simulations with varying parameter $\mu_3 \in [1.4, 4.2]$ and fixed values for other parameters: $\mu_1 = 1.85$, $\mu_2 = 0.54$, $\mu_4 = 0.5$, $\mu_5 = 0.75$, $\mu_6 = 0.85$, $\mu_7 = 0.8$, $\mu_8 = 0.95$, $\mu_9 = 0.01$, $\vartheta = 0.2$, and initial states $(x(0), y(0), z(0), w(0)) = (1, 2, 2, 0.01)$. The results of these simulations are presented in Figure 4.22 as bifurcation diagrams.

The introduction of memductance with logarithmic functions in the system has a noticeable impact, as seen in the formation of multiple periodic windows after an initial chaotic state. The chaotic nature of the system (4.14) is depicted in Figure 4.23, while the transition phases are illustrated in Figures 4.24 and 4.25. These plots provide insights into the dynamics of the system and demonstrate the effects of the logarithmic memductance on its behavior.

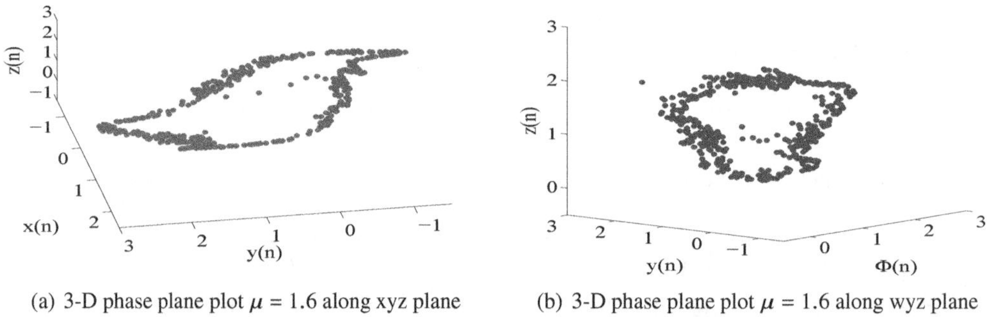

(a) 3-D phase plane plot μ = 1.6 along xyz plane

(b) 3-D phase plane plot μ = 1.6 along wyz plane

Fig. 4.23: 3-D phase plane plots of system (4.3) corresponding to Figure 4.22.

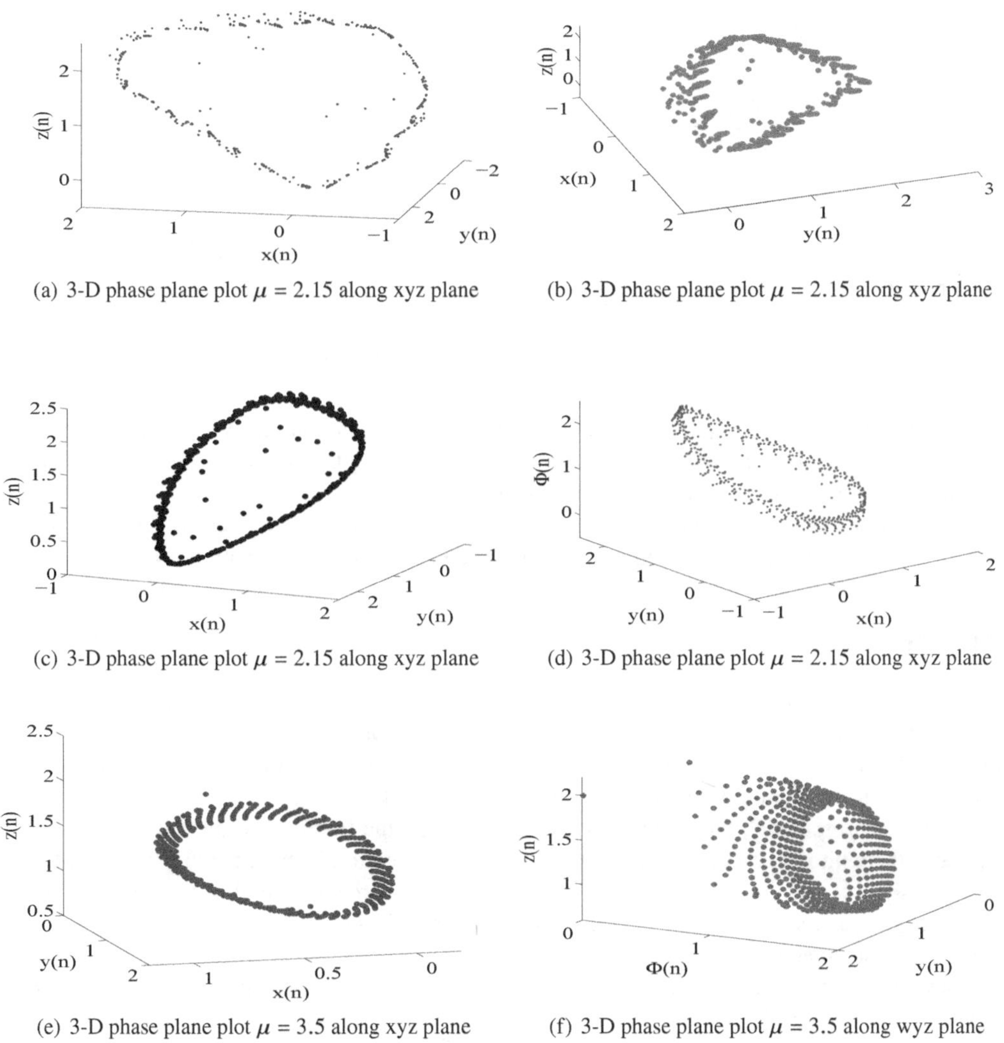

(a) 3-D phase plane plot $\mu = 2.15$ along xyz plane

(b) 3-D phase plane plot $\mu = 2.15$ along xyz plane

(c) 3-D phase plane plot $\mu = 2.15$ along xyz plane

(d) 3-D phase plane plot $\mu = 2.15$ along xyz plane

(e) 3-D phase plane plot $\mu = 3.5$ along xyz plane

(f) 3-D phase plane plot $\mu = 3.5$ along wyz plane

Fig. 4.24: 3-D phase plane plots of system (4.3) corresponding to Figure 4.22.

4.3.4 Discussion

This section focuses on the application of memristors in neuroscience using neuron models constructed with fractional difference operators. The goal is to enhance our understanding of the human brain and improve information transfer in various fields. Neuron models aim to replicate the behavior of biological neurons, and the incorporation of memristor elements allows for the theoretical representation of the neurons' ability to remember past pulses.

In this section, different flux-controlled memristive elements, such as nonlinear cubic, periodic, and exponential types, are introduced and their impact on the dynamical nature of the system is discussed. The analysis provides an interesting perspective on how variations in memristor elements influence the system's

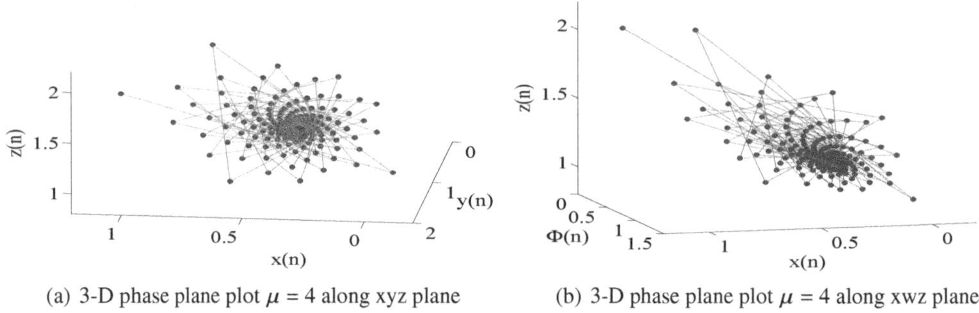

(a) 3-D phase plane plot $\mu = 4$ along xyz plane

(b) 3-D phase plane plot $\mu = 4$ along xwz plane

Fig. 4.25: 3-D phase plane plots of system (4.3) corresponding to Figure 4.22.

behavior. The bifurcation diagrams presented in this section demonstrate that the range of transition varies, with periodic elements exhibiting an extended phase. Additionally, the introduction of logarithmic memristor elements leads to the formation of multiple periodic windows.

Overall, this analysis sheds light on the relationship between memristors and neuroscience, showcasing their potential for advancing our understanding of the brain and improving information processing.

4.4 Summary

The chapter focuses on the discrete-time fractional order form of the KTz neuron model and explores its qualitative properties such as stability and chaos. The impact of fractional order is investigated, and time series plots are used to explain the complex dynamics exhibited by the variables. Coexisting attractors are also examined through coexisting bifurcations at different initial states.

To enhance the neuron model, the chapter introduces discrete-time fractional order memristive elements and incorporates them into the model. Bifurcation diagrams are used to illustrate the unpredictability of the system variables, showcasing the effect of the memristive elements. Three types of flux-controlled memristive elements are introduced, and their impact on the system dynamics is depicted through bifurcation diagrams accompanied by phase plane portraits.

By incorporating memristive elements into the neuron model, the chapter demonstrates an increased accuracy in modeling biological neurons, further enhancing our understanding of their behavior and dynamics.

References

[1] Jan Cermak, Istvan Gyori and Ludek Nechvatal. On explicit stability conditions for a linear fractional difference system. Fractional Calculus and Applied Analysis 18 (2015): 651–672.

[2] Adel Ouannas, Amina Aicha Khennaoui, Shaher Momani, Giuseppe Grassi and Viet-Thanh Pham. Chaos and control of a three-dimensional fractional order discrete-time system with no equilibrium and its synchronization. AIP Advances 10, no. 4 (2020): 045310.

[3] Guo-Cheng Wu and Dumitru Baleanu. Jacobian matrix algorithm for Lyapunov exponents of the discrete fractional maps. Communications in Nonlinear Science and Numerical Simulation 22, no. 1-3 (2015): 95–100.

[4] Alan L. Hodgkin and Andrew F. Huxley. A quantitative description of membrane current and its application to conduction and excitation in nerve. The Journal of Physiology 117, no. 4 (1952): 500.

[5] Von Neumann, J. and R. Kurzweil. The Computer and the Brain. Yale University Press (2012).

[6] Norbert Wiener. Cybernetics or Control and Communication in the Animal and the Machine. MIT Press (2019).

[7] Brian N. Lundstrom, Matthew H. Higgs, William J. Spain and Adrienne L. Fairhall. Fractional differentiation by neocortical pyramidal neurons. Nature Neuroscience 11, no. 11 (2008): 1335–1342.

[8] Osame Kinouchi and Marcelo H. R. Tragtenberg. Modeling neurons by simple maps. International Journal of Bifurcation and Chaos 6, no. 12a (1996): 2343–2360.

[9] Warren S. McCulloch and Walter Pitts. A logical calculus of the ideas immanent in nervous activity. The Bulletin of Mathematical Biophysics 5, no. 4 (1943): 115–133.

[10] Bharathwaj Muthuswamy. Implementing memristor based chaotic circuits. International Journal of Bifurcation and Chaos 20, no. 05 (2010): 1335–1350.

[11] Wei Liu, Fa-Qiang Wang and Xi-Kui Ma. Exponential flux-controlled memristor model and its floating emulator. Chinese Physics B 24, no. 11 (2015): 118401.

[12] Huagan Wu, Yi Ye, Mo Chen, Quan Xu and Bocheng Bao. Periodically switched memristor initial boosting behaviors in memristive hypogenetic jerk system. IEEE Access 7 (2019): 145022–145029.

[13] Quan Xu, Tong Liu, Cheng-Tao Feng, Han Bao, Hua-Gan Wu and Bo-Cheng Bao. Continuous non-autonomous memristive Rulkov model with extreme multistability. Chinese Physics B 30, no. 12 (2021): 128702.

[14] Ronilson Rocha, Jothimurugan Ruthiramoorthy and Thamilmaran Kathamuthu. Memristive oscillator based on Chua's circuit: Stability analysis and hidden dynamics. Nonlinear Dynamics 88, no. 4 (2017): 2577–2587.

[15] Malik, S. A. and A. H. Mir. FPGA realization of fractional order neuron. Applied Mathematical Modelling 81 (2020): 372–385.

[16] Dong Jun, Zhang Guang-Jun, Xie Yong, Yao Hong and Wang Jue. Dynamic behavior analysis of fractional-order Hindmarsh-Rose neuronal model. Cognitive Neurodynamics 8, no. 2 (2014): 167–175.

[17] Moaddy, K., Ahmed G. Radwan, Khaled N. Salama, Shaher Momani and Ishak Hashim. The fractional-order modeling and synchronization of electrically coupled neuron systems. Computers & Mathematics with Applications 64, no. 10 (2012): 3329–3339.

[18] Min Shi and Zaihua Wang. Abundant bursting patterns of a fractional-order Morris-Lecar neuron model. Communications in Nonlinear Science and Numerical Simulation 19, no. 6 (2014): 1956–1969.

[19] Argha Mondal, Sanjeev Kumar Sharma, Ranjit Kumar Upadhyay and Arnab Mondal. Firing activities of a fractional-order FitzHugh-Rinzel bursting neuron model and its coupled dynamics. Scientific Reports 9, no. 1 (2019): 1–11.

[20] Zhao Yao, Kehui Sun and Shaobo He. Firing patterns in a fractional-order FithzHugh-Nagumo neuron model. Nonlinear Dynamics 110, no. 2 (2022): 1807–1822.

[21] Leon Chua. Memristor-the missing circuit element. IEEE Transactions on Circuit Theory 18, no. 5 (1971): 507–519.

[22] Mohammad Saeed Feali, Arash Ahmadi and Mohsen Hayati. Implementation of adaptive neuron based on memristor and memcapacitor emulators. Neurocomputing 309 (2018): 157–167.

[23] Yunus Babacan, Firat Kacar and Koray Gurkan. A spiking and bursting neuron circuit based on memristor. Neurocomputing 203 (2016): 86–91.

[24] Fouda, M. E. and A. G. Radwan. On the fractional-order memristor model. Journal of Fractional Calculus and Applications 4, no. 1 (2013): 1–7.

[25] Yi-Fei Pu and Xiao Yuan. Fracmemristor: Fractional-order memristor. IEEE Access 4 (2016): 1872–1888.

[26] Karthikeyan Rajagopal, Anitha Karthikeyan, Sajad Jafari, Fatemeh Parastesh, Christos Volos and Iqtadar Hussain. Wave propagation and spiral wave formation in a Hindmarsh-Rose neuron model with fractional-order threshold memristor synapse. International Journal of Modern Physics B 34, no. 17 (2020): 2050157.

[27] Yajuan Yu, Min Shi, Huiyan Kang, Mo Chen and Bocheng Bao. Hidden dynamics in a fractional-order memristive Hindmarsh-Rose model. Nonlinear Dynamics 100, no. 1 (2020): 891–906.

Chapter 5

Discrete Fractional Model of Tumor and Effector Cells Interaction with Chemotherapy

This chapter aims to investigate the dynamic nature of a two-dimensional biological interaction model between tumor cells and effector cells, considering the effects of chemotherapy. The model incorporates competition between the two cell populations, as well as a fractional kill rate for the tumor cells to realistically represent their decline when targeted by activated immune cells. An activation rate is introduced for the effector cells, which is based on the fractional kill rate of the tumor cells. The chaotic behavior of the system is analyzed under both commensurate and incommensurate fractional order cases, and a comparison is made between the two. Bifurcation diagrams are used as a tool to study chaos in the system, and the breakdown of bifurcations is examined through time-varying plots of the state variables that describe the compartments. Furthermore, the study extends the two-dimensional model to a three-dimensional one by incorporating the concentration of chemotherapy drugs. The effects of the drugs are discussed specifically in the context of the incommensurate order case, considering two scenarios: constant rate of tumor kills and exponential kill rate. The analysis presents interesting findings regarding the dynamics of the system under varying drug inputs.

5.1 Introduction

Researchers extensively investigate the spread of diseases and develop strategies to control them in order to ensure the long-term survival of humanity. The field of medicine has made significant advancements, enabling the identification and study of new diseases and their causes. An alarming health concern involves the abnormal growth of cells or tissues in organisms, commonly referred to as tumors. These tumors can be categorized as either malignant or benign. Malignant tumors consist of cancerous cells that have the propensity to metastasize and invade healthy cells throughout the body. On the other hand, benign tumors, although less severe than malignant ones, do not typically spread or invade other body parts. Nonetheless, if left untreated, benign tumors can lead to fatal consequences. Therefore, timely detection and treatment of tumor cells are critical for effective management and complete recovery.

Cancer ranks as one of the leading causes of death worldwide, second only to cardiovascular diseases. Shockingly, cancer is responsible for nearly ten million human deaths, accounting for approximately one in six recorded deaths. The escalating fatality rate due to cancer underscores the urgent need for advancements in the field of oncology and the development of cancer vaccines.

5.1.1 Cancer Research

Cancer can be defined simply as the uncontrolled division and spread of abnormal cells, resulting in the formation of a mass known as a tumor. These cancerous cells have the ability to break away from the original tumor and spread to other parts of the body. The complexity of nature is exemplified by the human body and its intricate functioning. Within the human body, the immune system plays a vital role in identifying foreign objects and taking appropriate action to eliminate them. It is specifically designed to defend against infections and diseases. The immune system also acts to prevent the formation and growth of cancerous cells in the body, eradicating them in the early stages. As a result, individuals with compromised immune systems have a higher risk of developing cancer.

The study of tumors is not a recent area of interest but has been a topic of investigation since ancient times. This can be observed through the use of the term "carcinoma" by Hippocrates around 400 BC. While there were notable research developments during the 18th century, it was in the mid-19th century that the study of cancer intensified with the formulation of the tumor cell theory by Rudolf Virchow. The significance given to cancer research by various countries is evident from the establishment of cancer research societies, such as those in the UK in 1902 and the US in 1913. Research focus in the 20th century revolved around the understanding of tumor metastasis to other parts of the body. In the 21st century, advancements in technology have propelled cancer research to new levels of investigation and understanding.

Immune System and Tumors:
Detecting and responding to tumor cells involves the production of tumor antigens, activation of immune cells, and direct attacks on tumor cells through cytokine release. Immune cells are vital in recognizing tumor antigens and taking action to inhibit tumor growth. Effector cells, including T-cells, killer cells, and macrophages, play a crucial role in identifying and eliminating abnormal cells. These cells form an essential part of the human immune system and provide immunity against tumor cells.

1. Cytotoxic T lymphocytes (CTLs) - Recognize antigens on target cells.
2. Natural killer cells (NK) - Detect tumor cells without antigens.
3. Natural Killer T cells (NKT) - Activator of immune cells (innate and adaptive).
4. Macrophages
   - Tumor associated macrophages (TAM) - TAM-1(M1)[Kill tumor cells] and TAM -2 (M2) [For tumor tolerance].
5. Lymphokines - Stimulation of the growth of immune cells.
6. Dendritic cells - Initiate tumor specific responses.

Despite the involvement of various cells in combating tumor growth, deficiencies in immune response contribute to the development of tumors. Factors like the suppression of immune activity by certain T-cells, tumor cells' ability to evade immune detection, external agents (such as viruses like HIV or chemicals such as drugs) that weaken the immune barrier, and the immune system's incapability to effectively activate cytokines all play a role in tumor formation.

In response to these challenges, the field of cancer research has focused on identifying methods to hinder tumor cell growth. These methods aim to control tumor growth and address the possibility of cancer recurrence. Consequently, studying cancer cells presents a significant challenge for researchers in medicine, biochemistry, biotechnology, and molecular biology. Oncology research and advancements have led to improved cancer detection and treatment strategies. Notably, the ability to predict cancer occurrence based on test results underscores the progress achieved in oncological research. A detailed illustration on tumor and immune cell interactions can be obtained by the readers from [6, 7, 9, 11].

5.2 Mathematical Models of Tumor-Immune System

The recent success in immunotherapies for cancer has highlighted the importance of understanding the interactions between tumor cells and immune cells. Mathematical models provide a suitable framework for studying these relationships and allow for the incorporation of external factors that influence tumor growth and immune cell suppression. By using mathematical analysis, we can gain valuable insights into the behavior and dynamics of tumor cells, providing a clear direction for practical approaches to dealing with cancer while also being cost-efficient.

While it is unrealistic to include all factors involved in the tumor-immune cell interaction within a mathematical model, simple models may not capture the complex nature of tumor cell behavior and dynamics. Therefore, it is crucial to develop biologically realistic models that capture important characteristics of the cell compartments under study. One such model, developed by Kirschner and Panetta in 1998 [3], focuses on the interaction between effector cells, tumor cells, and IL-2 cells.

$$\begin{cases} \dfrac{dE}{dt} = cT - \mu_2 E + \dfrac{p_1 E I_L}{g_1 + I_L} + s_1 \\ \dfrac{dT}{dt} = r_2(T)T - \dfrac{aET}{g_2 + T} \\ \dfrac{dI_L}{dt} = \dfrac{p_2 ET}{g_3 + T} - \mu_3 I_L + s_2. \end{cases} \tag{5.1}$$

Mathematical modelling of the interaction with fractional kill rate was proposed by Pillis and Radunskaya in [4] given by

$$\begin{cases} \dfrac{dT}{dt} = aT(1 - bT) - cNT - D \\ \dfrac{dN}{dt} = \sigma - fN + \dfrac{gT^2 N}{h + T^2} - pNT \\ \dfrac{dL}{dt} = -mL + \dfrac{jD^2 L}{k + D^2} L - qLT + rNT \end{cases} \tag{5.2}$$

The chapter discusses various mathematical models that explore the interactions between tumor and immune systems, taking into account factors such as memory and hereditary characteristics. Fractional order derivatives are employed in these models to capture the complex dynamics of the systems. Ahmed et al. studied tumor-immune interactions with fractional derivatives in their work [14], while Rihan et al. investigated the dynamics of a tumor-immune system with delay in [15]. The qualitative behavior and stability of tumor-immune systems with fractional derivatives were discussed in [16], and bifurcation analysis of a chemotherapy model with fractional derivatives was performed in [28]. The influence of cancer cells on the immune system was illustrated in [17], and numerical modeling of these interactions was carried out in [18]. Conformable fractional derivative models were presented in [20], and contributions to fractional order models were made in [19] and [21].

While significant progress has been made in the literature regarding continuous fractional derivatives, less attention has been given to discrete fractional operators. Chaotic behavior of immune and tumor cells using discrete fractional operators was presented in [22], and Balci et al. studied discrete conformable fractional derivative models in [23]. To contribute strong results and understand the applicability of discrete fractional operators, this chapter focuses on a tumor interaction model with effector cells, utilizing the Caputo fractional difference operator.

5.2.1 Discrete Fractional Tumor and Effector Cells Interaction Model

Two interacting cell populations are considered for study in this chapter. They are as under:

- Tumor cells at time n denoted by $x(n)$,
- Effector cells population at time n denoted by $y(n)$.

The following assumptions are considered for the development and dynamical analysis of the model:

Q_1 The tumor cells grow logistically in the absence of immune response (i.e., the growth of the tumor cells is deterministic).
Q_2 External factors, including environmental factors like temperature and nutrients, are assumed to be constant.
Q_3 The behaviour of all the tumor cells is assumed to be similar.
Q_4 The effector cells are considered to always be present as a result of innate immunity.
Q_5 Effector cell populations are assumed to kill tumor cells upon interaction.
Q_6 The environment around tumor cells is assumed to be non-immunosuppressive.

In this chapter, certain assumptions are made to impose restrictions on our study. Assumption Q_2 is considered to focus on factors that can be evaluated, as some factors related to the tumor-immune interaction are not yet fully explored. Assumption Q_3 limits the discussion to the heterogeneity of the tumor cell population, allowing for a general analysis of the interaction rather than specific cases or situations. Assumptions Q_5 and Q_6 are introduced to highlight the fact that our focus in this chapter is on effector cells, which are a group of immune cells that act against tumor cells. It is important to note that some of these cells have the ability to inhibit tumor growth, while others can stimulate it. This complexity adds challenges to understanding the dynamics of the populations involved.

Overall, these assumptions serve to provide a framework for our study and guide the analysis of the tumor-immune interaction in this chapter.

Tumor Cells:
The change in tumor cell population can be modelled with logistic growth, and the decrease in tumor cell populations can be represented with fractional kill rate and competition with effector cells. The equation representation can then be given by

$$\frac{dx}{dt} = \theta x(1-x) - cxy - b\Psi(x,y)x, \tag{5.3}$$

where θ is the rate of growth of tumor cells, c is competition rate, $\Psi(x,y) = \dfrac{d(\frac{y}{x})^p}{s+(\frac{y}{x})^p}$

Effector Cells:
Effector cells encompass various types of immune cells that actively combat tumor cell growth. These cells are consistently present in our body as part of innate immunity. Instead of incorporating a growth rate, the effector cell population is described by an activation term, which depends on the encounter with tumor cells, their fractional kill rate, and the subsequent population decline resulting from the interaction. Therefore, the equation that represents the effector cell population can be expressed as follows:

$$\frac{dy}{dt} = -gx - wxy - u\frac{(\Psi(x,y))^2x^2y}{h+(\Psi(x,y))^2x^2}, \tag{5.4}$$

where g is the natural death rate, w is the rate of decline due to competition, and u is the rate of activation. Combining equations (5.3) and (5.4), we arrive at the following system of tumor-immune cell interaction models:

$$\begin{cases} \dfrac{dx}{dn} = & \theta x(1-x) - cxy - b\Psi(x,y)x, \\ \dfrac{dy}{dn} = & -gx - wxy - u\dfrac{(\Psi(x,y))^2x^2y}{h+(\Psi(x,y))^2x^2}. \end{cases} \tag{5.5}$$

This chapter considers the discrete–time fractional order model of tumor cells and effector immune cells with a fractional kill rate of tumor cells, so the system (5.5) takes the following form:

$$\begin{cases} \Delta^{\vartheta}x(n) = & \theta x(n-1+\vartheta)(1-x(n-1+\vartheta)) - bx(n-1+\vartheta)\Psi(x(n-1+\vartheta), y(n-1+\vartheta)) \\ & -cx(n-1+\vartheta)y(n-1+\vartheta), \\ \Delta^{\vartheta}y(n) = & -gx(n-1+\vartheta) - wx(n-1+\vartheta)y(n-1+\vartheta) \\ & +u\dfrac{(\Psi(x(n-1+\vartheta), y(n-1+\vartheta)))^2x(n-1+\vartheta)^2}{h+(\Psi(x(n-1+\vartheta), y(n-1+\vartheta)))^2x(n-1+\vartheta)^2}y(n-1+\vartheta), \end{cases} \tag{5.6}$$

where Δ^{ϑ} is the Caputo difference operator with $0 < \vartheta \leq 1$, $\Psi(x(n), y(n)) = \dfrac{dy(n)^p}{sx(n)^p + y(n)^p}$. For further analysis, the above considered system is converted to the numerical form [2] as follows:

$$\begin{cases} x(n) = & x(0) + \dfrac{1}{\Gamma(\vartheta)}\sum_{k=1}^{n}\dfrac{\Gamma(n-k+\vartheta)}{\Gamma(n-k+1)}\Bigg[\theta x(k-1)(1-x(k-1)) - bx(k-1)\Psi(x(k-1), y(k-1)) \\ & -cx(k-1)y(k-1)\Bigg], \\ y(n) = & y(0) + \dfrac{1}{\Gamma(\vartheta)}\sum_{k=1}^{n}\dfrac{\Gamma(n-k+\vartheta)}{\Gamma(n-k+1)}\Bigg[-gx(k-1) - wx(k-1)y(k-1) \\ & +u\dfrac{(\Psi(x(k-1), y(k-1)))^2x(k-1)^2}{h+(\Psi(x(k-1), y(k-1)))^2x(k-1)^2}y(k-1)\Bigg], \end{cases} \tag{5.7}$$

where $n = 1, 2, 3\cdots$.

5.3 Commensurate Order System

This section of the chapter concentrates on investigating the stability and chaotic behavior exhibited by the model (5.6) with identical fractional orders. Though the interest lies in understanding the chaotic dynamics of the considered 2-D model, stability analysis is performed for comparison with the system stability when non-identical orders are considered in the proceeding sections.

5.3.1 Stability Analysis

The stability analysis is performed numerically by evaluating the eigenvalues of the system obtained from the Jacobian matrix, and the results are verified based on the following theorem:

Theorem 5.1 *[1]Let $\vartheta \in (0,1)$ and $\Psi \in \mathbb{R}^{k\times k}$. Let*

$$B^{\vartheta} = \left\{\lambda \in \mathbb{C} : |\lambda| < \left(2\cos\left(\frac{|arg\lambda| - \pi}{2-\vartheta}\right)\right)^{\vartheta} \; and \; |arg\lambda| > \frac{\vartheta\pi}{2}\right\}.$$

If all the eigenvalues of Ψ lie in B^{ϑ}, then the system (4.3) is asymptotically stable. The system (5.6) is unstable if $\lambda \in \mathbb{C}\backslash cl(B^{\vartheta})$ for an eigenvalue of Ψ.

Let the parameters of the systems take the following values: $\theta = 1.5$, $b = 0.8345$, $c = 0.14$, $d = 0.6$, $g = 0.3$, $u = 0.5$, $w = 0.05$, $h = 0.225$, $p = 0.5$, $s = 0.6$ and the fractional order $\vartheta = 0.9$. The equilibrium states of the system are:

1. Axial equilibrium point $E_0 = (1, 0)$.
2. Interior equilibrium point $E_1 = (0.3899287251, 3.556953880)$.
3. Interior equilibrium point $E_2 = (0.6597507726, 1.180124509)$.

The results are depicted for the interior equilibrium points E_1 and E_2.

Interior Equilibrium Point E_1:
The Jacobian matrix of the system (5.6) at E_1 is

$$J(E_1) = \begin{bmatrix} -0.4952910135 & -0.05837934437 \\ 2.947696680 & -0.0749428651 \end{bmatrix}, \tag{5.8}$$

corresponding eigenvalues are conjugate complex numbers $\eta_{1,2} = -0.2851169392 \pm i\,0.3576471139$. From Theorem 5.1, the argument of the eigen values are

$$argument(-0.2851169392 \pm i\,0.3576471139) = \pm 2.243828747$$

and the absolute values are $|\eta_{1,2}| = 0.4573872835$. Finally,

$$\left(2\cos\left(\frac{|arg(-0.2851169392 \pm i\,0.3576471139)| - \pi}{2-0.9}\right)\right)^{0.9} = 1.326136596. \tag{5.9}$$

On comparing the numerically evaluated values, we get

$$|\eta_{1,2}| < \left(2\cos\left(\frac{|arg(-0.2851169392 \pm i\,0.3576471139)| - \pi}{2-0.9}\right)\right)^{0.9}$$

and the arguments of the eigenvalues are clearly greater than $\frac{0.9\pi}{2} = 1.414285714$. The observations ensure the asymptotic stability of the system at the initerior equilibrium point E_1.

Interior Equilibrium Point E_2 :
We now shift our focus towards the other interior equilibrium state of the system (5.6). At E_2, the Jacobian matrix is

$$J(E_2) = \begin{bmatrix} -1.016056288 & -0.1222437776 \\ 0.7367071500 & 0.1186408752 \end{bmatrix}. \tag{5.10}$$

The eigenvalues are obtained as $\eta_1 = -0.9301914008$ and $\eta_2 = 0.03277598807$. From the direct observation of the eigenvalues, it can be concluded that the presence of a positive eigenvalue η_2 indicates the unstable nature of the system, which can be proved to be true with numerical evidence.

$$\begin{aligned} \left(2\cos\left(\frac{|arg(-0.9301914008)| - \pi}{2-0.9}\right)\right)^{0.9} &= 1.866064873, \\ \left(2\cos\left(\frac{|arg(0.03277598807)| - \pi}{2-0.9}\right)\right)^{0.9} &= -1.710419308 + i\,0.5557489219. \end{aligned} \tag{5.11}$$

The argument of η_1 is 3.141592654 and argument of η_2 is 0. The instability of the system becomes apparent when it is observed that the argument of one of the eigenvalues is zero. This violates the condition of having an argument value greater than $\frac{0.9\pi}{2} = 1.414285714$. As a result, the system described by Equation (5.6) is proven to be unstable at the equilibrium state E_2, as the condition from Theorem (5.1) is not fulfilled.

5.3.2 Chaotic Dynamics of Tumor and Effector Cells Interaction

Investigating the dynamic behavior of tumor cell and effector cell populations is crucial for the advancement of control strategies in the interdisciplinary field of medicine. Understanding how these populations interact and evolve over time provides valuable insights for developing effective control measures.

Bifurcation diagrams are simulated for $b = 3.125$, $c = 0.14$, $d = 0.16$, $g = 0.3$, $u = 0.3$, $w = 0.75$, $h = 0.2$, $p = 0.5$, $s = 0.16$ and the fractional order $\vartheta = 0.3$ with varying growth rate of the tumor cells θ over the range $(1.5, 2.7)$.

Analyzing the chaotic dynamics is essential for understanding the significant changes that can occur in a system. In the context of tumor cells and effector cells, the impact of tumor cells on the human body becomes increasingly significant as the tumor grows. Therefore, we introduce the growth rate of tumor cells, denoted as θ, as a bifurcation parameter to investigate the system's behavior. Increasing the growth rate of tumor cells leads to an escalation in the chaotic behavior of both populations, as depicted in Figure 5.1. Bifurcation diagrams are used to illustrate the system's behavior at different values of θ through time series plots. By examining these plots individually, we can observe the changes in each population.

Figures 5.2, 5.3, and 5.4 present the time series plots for various values of θ, namely $\theta = 1.8, 2.15, 2.38, 2.45, 2.55$, and 2.65. For small growth rates of tumor cells, both cell populations exhibit

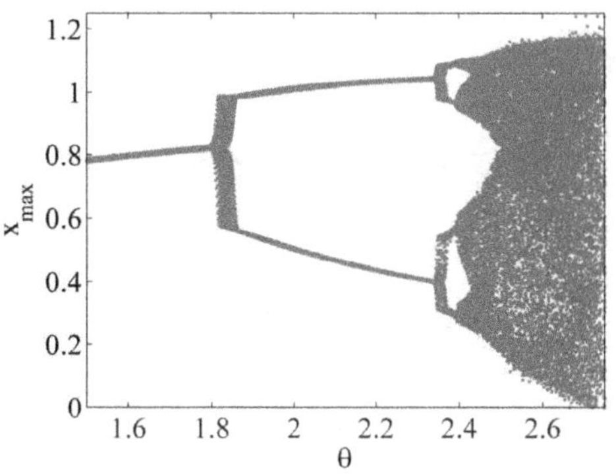

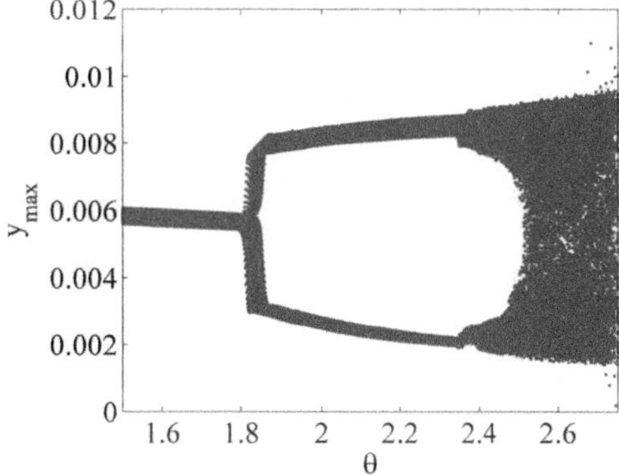

Fig. 5.1: Bifurcation diagrams of the system (5.6).

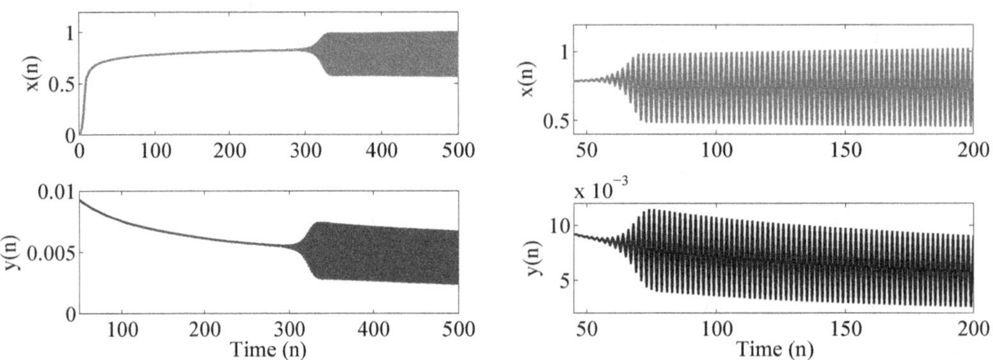

Fig. 5.2: Time series plots corresponding to bifurcation diagrams of the system (5.6) in Figure 5.1 for $\theta = 1.8, 2.15$.

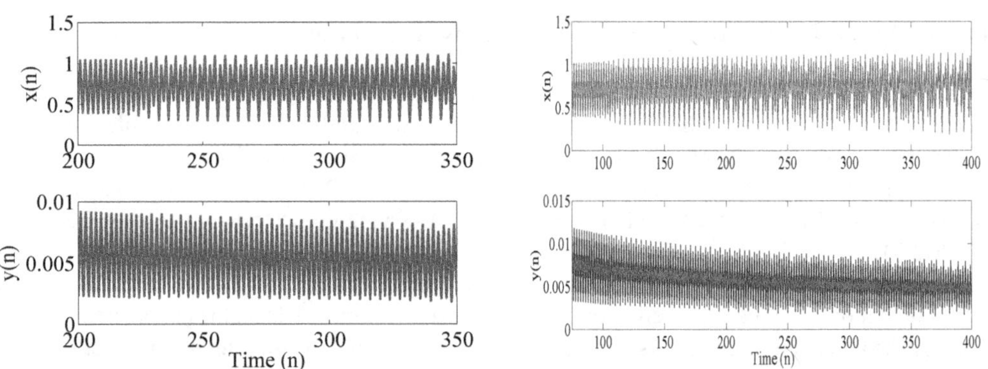

Fig. 5.3: Time series plots corresponding to bifurcation diagrams of the system (5.6) in Figure 5.1 for $\theta = 2.38, 2.45$.

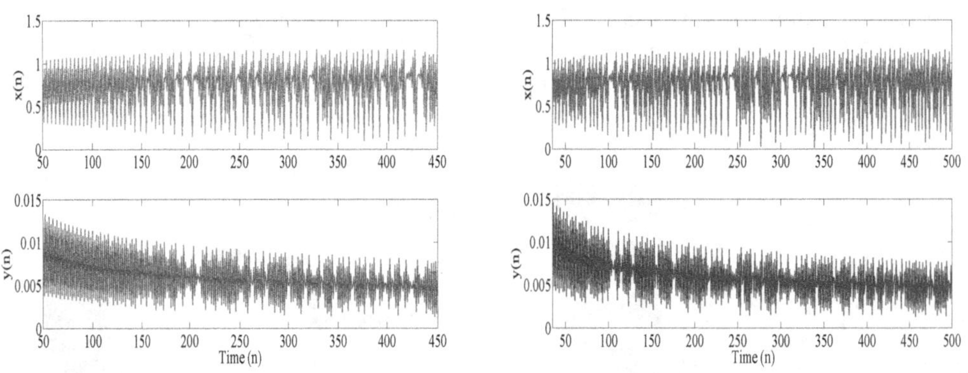

Fig. 5.4: Time series plots corresponding to bifurcation diagrams of the system (5.6) in Figure 5.1 for $\theta = 2.55, 2.65$.

relatively stable behavior. However, as the tumor population reaches a certain level of growth, the influence of effector cells on the tumor cells intensifies, resulting in oscillatory behavior of the populations. When $\theta > 2.45$, the system initially displays periodic oscillations, which eventually transition's into aperiodic oscillations over time. This aperiodic behavior of the cells is clearly observed in Figure 5.4 when $\theta = 2.65$. Therefore, when considering cell populations with identical orders, the growth rate of tumor cells emerges as a crucial factor that induces random behavior in the cell populations.

5.4 Tumor and Effector Cells Interaction Model with Incommensurate Order

In this section of the chapter, we introduce the concept of studying the interaction model between tumor and immune cells with non-identical fractional orders. Biologically, no two cells exhibit the same behavior or characteristics. Each type of cell has different constituents that make them suitable for specific roles within biological mechanisms. Therefore, it becomes essential to identify and explore the unique properties of these cells.

Constructing a mathematical model with identical orders may not be suitable for accurately describing the diverse features of biological cells, which exhibit different behaviors under various factors. To overcome this limitation, we propose to investigate the chaotic dynamics displayed by the interaction model between tumor and immune cells with non-identical fractional orders [2]. By introducing non-identical orders, we aim to uncover unknown properties of the cells, which we will compare with the identical order model discussed in the previous section.

First, we consider the incommensurate order form of the system described by Equation (5.6).

$$\begin{cases} \Delta^{\vartheta_1}x(n) = & \theta x(n-1+\vartheta_1)(1-x(n-1+\vartheta_1)) - bx(n-1+\vartheta_1)\Psi(x(n-1+\vartheta_1), y(n-1+\vartheta_1)) \\ & -cx(n-1+\vartheta_1)y(n-1+\vartheta_1), \\ \Delta^{\vartheta_2}y(n) = & -gx(n-1+\vartheta_2) - wx(n-1+\vartheta_2)y(n-1+\vartheta_2) \\ & +u\dfrac{(\Psi(x(n-1+\vartheta_2), y(n-1+\vartheta_2)))^2 x(n-1+\vartheta_2)^2}{h+(\Psi(x(n-1+\vartheta_2), y(n-1+\vartheta_2)))^2 x(n-1+\vartheta_2)^2} y(n-1+\vartheta_2), \end{cases} \tag{5.12}$$

where $\Delta^{\vartheta_i}, i = 1, 2$ is the Caputo difference operator with $0 < \vartheta_{1,2} \leq 1$, $\Psi(x(n), y(n)) = \dfrac{dy(n)^p}{sx(n)^p + y(n)^p}$. For further analysis, the system described above is transformed into a numerical form, as presented in [2], which can be expressed as follows:

$$\begin{cases} x(n) = & x(0) + \dfrac{1}{\Gamma(\vartheta_1)} \sum_{k=1}^{n} \dfrac{\Gamma(n-k+\vartheta_1)}{\Gamma(n-k+1)} \Bigg[\theta x(k-1)(1-x(k-1)) - bx(k-1)\Psi(x(k-1), y(k-1)) \\ & -cx(k-1)y(k-1) \Bigg], \\ y(n) = & y(0) + \dfrac{1}{\Gamma(\vartheta_2)} \sum_{k=1}^{n} \dfrac{\Gamma(n-k+\vartheta_2)}{\Gamma(n-k+1)} \Bigg[-gx(k-1) - wx(k-1)y(k-1) \\ & +u\dfrac{(\Psi(x(k-1), y(k-1)))^2 x(k-1)^2}{h+(\Psi(x(k-1), y(k-1)))^2 x(k-1)^2} y(k-1) \Bigg], \end{cases} \tag{5.13}$$

where $n = 1, 2, 3\cdots$.

5.4.1 Dynamical Analysis of Incommensurate Order System (5.12)

The investigation focuses on analyzing the stability properties of the system and exploring its chaotic behavior through the use of bifurcation diagrams. The study primarily revolves around the interior equilibrium points, aiming to depict the dynamic nature of the two cell populations.

To obtain stability results for incommensurate order systems, the following theorem is utilized:

Theorem 5.2 *[5] Let* $\alpha_\kappa = \frac{\omega_\kappa}{\beta_\kappa} \in (0,1)$, *where* $\kappa = 1, 2, 3, \cdots, m$, $\omega_\kappa, \beta_\kappa > 0$ *are positive integers, let the LCM of the denominators be* Φ. *The trivial solution of the system (5.12) with initial condition* $x(0) = x_0$ *is asymptotically stable if the equation*

$$det\left(diag\left(\eta^{\Phi\alpha_1}, \eta^{\Phi\alpha_2}, \dots, \eta^{\Phi\alpha_m}\right) - (1 - \eta^{\Phi} J)\right) = 0, \tag{5.14}$$

has all the roots inside the set $\mathbb{C}\backslash B_1^{\vartheta}$, *where* $\vartheta = \frac{1}{\Phi}$ *and*

$$B_1^{\vartheta} = \left\{\lambda \in \mathbb{C} : |\lambda| \le \left|\left(2\cos\frac{|arg\lambda|}{\vartheta}\right)^{\vartheta}\right| \text{ and } |arg\lambda| \le \frac{\vartheta\pi}{2}\right\}.$$

Equilibrium States and Stability:
The equilibrium states of the system (5.12) are numerically calculated for the following parameter values: $\theta = 1.5$, $b = 3.125$, $c = 0.14$, $d = 0.16$, $g = 0.3$, $w = 0.1$, $u = 0.5$, $h = 0.225$, $p = 0.5$, $s = 0.16$ and the fractional orders $\vartheta_1 = 0.8$ and $\vartheta_2 = 0.1$. The equilibrium states of the system are

1. Axial equilibrium point $E_0 = (1, 0)$.
2. Interior equilibrium point $E_1 = (0.4097388865, 2.953659001)$.
3. Interior equilibrium point $E_2 = (0.5640872928, 1.425639977)$.

Interior Equilibrium Point E_1:
At E_1, the Jacobian matrix is given by

$$J(E_2) = \begin{bmatrix} -0.5140787198 & -0.06254097699 \\ 2.248594006 & -0.0807895320 \end{bmatrix}. \tag{5.15}$$

Using the equation (5.14), we arrive at the following:

$$\begin{aligned} \eta^9 &- 0.0807895320\eta^{18} + 0.0807895320\eta^8 - 0.5140787198\eta^{11} \\ &+ 0.1821614452\eta^{20} - 0.3643228904\eta^{10} + 0.5140787198\eta + 0.1821614452 = 0. \end{aligned} \tag{5.16}$$

It is clear that
$|\eta_l| > \left|(2\cos(10\,|arg\eta_l|))^{0.1}\right|$ for $l \in \{1, 2, 3, 5, 7, 8, 11, 13, 14, 16, 18, 19, 20\}$ and $|arg\eta_l| > \frac{\pi}{20}$ for $l \in \{2, 3, 4, 5, 6, 7, 8, 9, 10, 11, 12, 13, 14, 15, 16, 17, 18, 19\}$. From Theorem 5.2 $\eta_l \in \mathbb{C}\backslash B_1^{\vartheta}$ for $1 \le l \le 20$, ensures the asymptotic stability of the system at the interior equilibrium point E_1.

Interior Equilibrium Point E_2:
At E_2, the Jacobian matrix is given by

$$J(E_2) = \begin{bmatrix} -0.8253610382 & -0.08719031049 \\ 2.534839418 & 0.2655290647 \end{bmatrix}. \tag{5.17}$$

Table 5.1: Roots of the equation (5.16).

η_1 = 1.12398+i 0.16848	η_{11} = -1.17720
η_2 = 0.85974 +i 0.45000	η_{12}= -0.79687-i 0.37488
η_3 = 0.87803+ i 0.81209	η_{13} = -1.02076-i 0.57530
η_4 = 0.33064+i 0.83886	η_{14} = -0.52958 -i 1.06935
η_5 = 0.22010+i 1.18246	η_{15}= -0.29957- i 0.82653
η_6 = -0.29957+i 0.82653	η_{16} = 0.22010-i 1.18246
η_7 = -0.52958+i 1.06935	η_{17} = 0.33064-i 0.83886
η_8 = -1.02076+ i 0.57530	η_{18} = 0.87803-i 0.81209
η_9 = -0.79687+i 0.37488	η_{19} = 0.85974-i 0.45000
η_{10} = -0.35420	η_{20} = 1.12398-i 0.16848

Using equation (5.14), we arrive at the following:

$$\begin{aligned}&\eta^9+0.265529064\eta^{18} - 0.2655290647\eta^8 - 0.8253610382\eta^{11}\\&\quad + 0.0018560914\eta^{20} - 0.0037121828\eta^{10} + 0.8253610382\eta + 0.0018560914 = 0.\end{aligned} \tag{5.18}$$

Table 5.2: Roots of the equation (5.18).

η_1 = 1.08865+i 0.17519	η_{11} = -1.09996
η_2 = 0.85218 +i 0.50202	η_{12}= -0.86731-i 0.43417
η_3 = 0.70121+ i 1.01676	η_{13} = -0.98378-i 0.66637
η_4 = 0.33044+i 0.87074	η_{14} = -0.31123 -i 0.85668
η_5 = 5.35335 $\times 10^{-7}$+i 11.96069	η_{15}= -0.25905- i 1.24902
η_6 = -0.25905+i 1.24902	η_{16} = 5.35335$\times 10^{-7}$-i 11.96069
η_7 = -0.31123+i 0.85668	η_{17} = 0.33044-i 0.87074
η_8 = -0.98378+ i 0.66637	η_{18} = 0.70121-i 1.01676
η_9 = -0.86731+i 0.43417	η_{19} = 0.85218-i 0.50202
η_{10} = -0.00224	η_{20} = 1.08865-i 0.17519

It is clear that
$|\eta_l| > \left|(2\cos(10\,|arg\eta_l|))^{0.1}\right|$ for $l \in \{1,3,5,6,8,9,11,12,13,15,16,18,20\}$ and
$|arg\eta_l| > \dfrac{\pi}{20}$ for $l \in \{1,2,3,4,5,6,7,8,9,10,11,12,13,14,15,16,17,18,19,20\}$. From Theorem 5.2 it is clear that $\eta_l \in \mathbb{C}\backslash B_1^{\vartheta}$ for $1 \le l \le 20$, which ensures the asymptotic stability of the system at the interior equilibrium point E_2.

In the case of the commensurate order system described by Equation (5.6), only one of the two interior equilibrium states achieves stability, while the other remains unstable. However, for a system with incommensurate orders, both interior equilibrium states become stable. This enhancement of the system's stability properties when the orders of the state variables are altered highlights the significance of studying non-identical fractional order systems. The introduction of this additional parameter provides an advantage in controlling the dynamics of the system.

5.4.2 Bifurcation Analysis

The chaotic behavior observed in biological systems has captured significant attention from researchers and scientists, as it reveals the inherent randomness of the system and provides insights into the factors contributing to this behavior. Developing control strategies based on these parameters can ensure the sustainability of biological populations.

In this analysis, specific values are assigned to the parameters: $b = 0.13889$, $c = 0.14$, $d = 0.36$, $g = 0.5$, $w = 0.475$, $u = 0.02$, $h = 0.72$, $p = 0.5$, and $s = 0.16$. The fractional orders are set as $\vartheta_1 = 0.7$ and $\vartheta_2 = 0.1$, while the growth rate parameter θ varies within the range of 1.3 to 2.67.

The presented bifurcation diagrams in Figure 5.5 exhibit a periodic doubling bifurcation. The system initially displays stable behavior, transition's to periodic oscillations, and eventually becomes chaotic. These transitional states are illustrated through time-varying plots in Figures 5.6, 5.7, and 5.8. For a growth rate of tumor cells of $\theta = 1.8$ and 2.15, the system demonstrates uniform oscillations, with the amplitude changing between these values. This amplitude change indicates how the cell populations shift from one state to another as the growth rate of tumor cells varies.

When $\theta = 2.3$, the system exhibits period-4 oscillations of both tumor and immune cells, as shown in Figure 5.7. However, these period-4 oscillations become aperiodic when $\theta = 2.5$. This pattern of non-periodic oscillations continues for all values of θ greater than 2.5, as depicted in Figure 5.8. It is observed that as the growth rate of tumor cells increases, the system becomes more chaotic. From the numerical simulations, it can be understood that controlling the growth rate of tumor cells is crucial for containing the populations of tumor and effector cells. Only by effectively controlling the tumor growth can the populations be managed and maintained. A comparison with the commensurate order system described by Equation (5.6) reveals that the system's chaotic behavior has improved when non-identical orders are assigned to the state variables.

5.4.3 Impact of Fractional Order on System Dynamics

To gain insights into the dynamics of the system under varying fractional orders, we consider fixing one fractional order as $\vartheta_2 = 0.1$ and varying the other order as $\vartheta = 0.3, 0.6, 0.9$. The behavioral changes exhibited by the system are presented through a sequence of bifurcation diagrams. The remaining parameter values are set as follows: $b = 0.13889$, $c = 0.14$, $d = 0.36$, $g = 0.5$, $w = 0.475$, $u = 0.02$, $h = 0.72$, $p = 0.5$, $s = 0.36$, and θ ranges from 1.5 to 3.

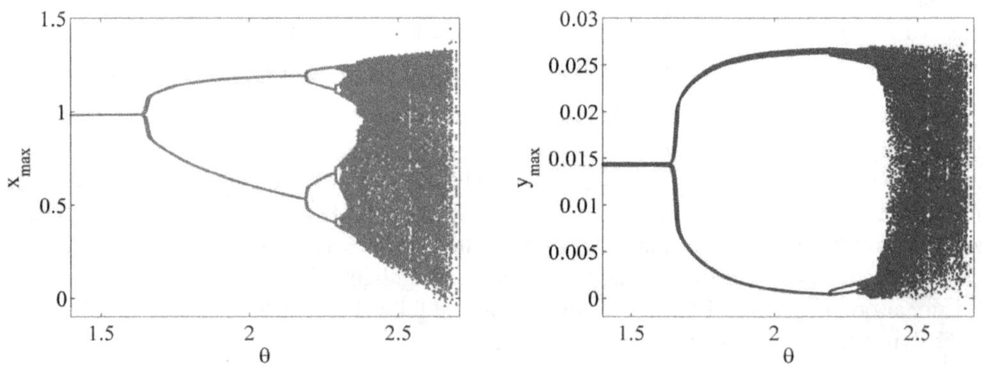

Fig. 5.5: Bifurcation diagrams of the system (5.12).

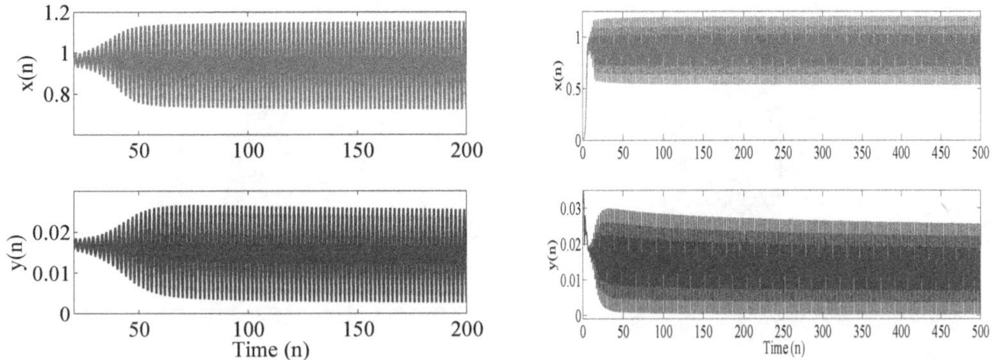

Fig. 5.6: Time series plots corresponding to bifurcation diagrams of the system (5.12) in Figure 5.5 for $\theta = 1.8, 2.15$.

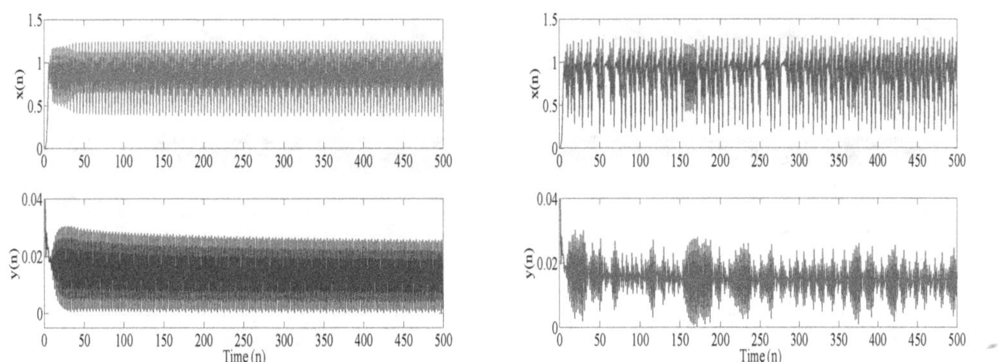

Fig. 5.7: Time series plots corresponding to bifurcation diagrams of the system (5.12) in Figure 5.5 for $\theta = 2.3, 2.5$.

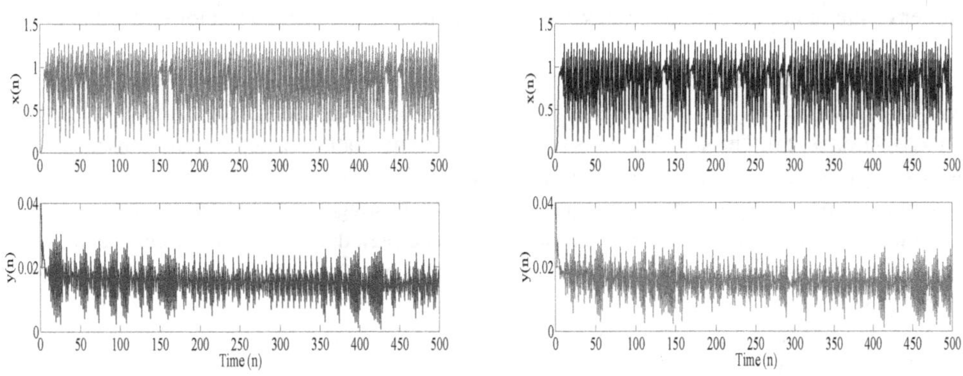

Fig. 5.8: Time series plots corresponding to bifurcation diagrams of the system (5.12) in Figure 5.5 for $\theta = 2.6, 2.65$.

The numerical simulations, as shown in Figure 5.9, depict a sequence of bifurcation diagrams with variations in two parameters, illustrating how the fractional order influences the dynamics of the cell populations. Unlike the previous section, where only the growth rate parameter was considered, here the focus is on the

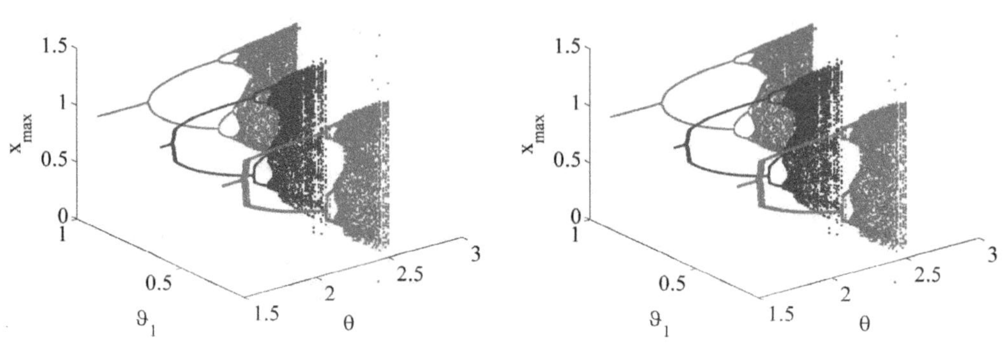

Fig. 5.9: Bifurcation diagrams of the system (5.12).

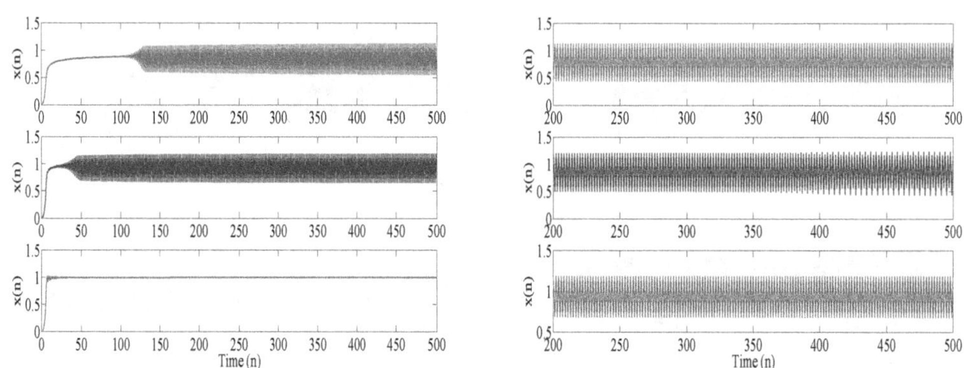

Fig. 5.10: Time series plots corresponding to bifurcation diagrams of the system (5.12) in Figure 5.9 for $\theta = 1.8, 2.15$.

fractional order of the tumor cell population, while the fractional order of the effector cells remains fixed at $\vartheta_2 = 0.1$.

Understanding the dynamics of tumor cells under different scenarios is crucial for gaining insights into their behavioral aspects. In this context, we consider fractional orders of the state variable as $\vartheta_1 = 0.3$ (red), 0.6 (blue), and 0.9 (magenta). The simultaneous time-varying behavior of the cell populations under different growth rates of the tumor cells is presented in Figures 5.10 and 5.11 for $\theta = 1.8, 2.15, 2.35$, and 2.5.

At the first glance, the bifurcation diagrams in Figure 5.9 might suggest that increasing the fractional order improves chaoticity. However, upon closer observation, it can be seen that for smaller orders such as $\vartheta_1 = 0.3$ and higher orders such as $\vartheta_1 = 0.9$, the system bifurcates at higher values of θ compared to the system with $\vartheta_1 = 0.6$. Furthermore, the system with $\vartheta_1 = 0.6$ enters chaotic regions before the other two. This discussion is supported by the time-varying plots in Figure 5.10, where the blue curve exhibits oscillations starting at an earlier value of n compared to the other two curves. The image to the right in the figure provides a glimpse of the changing oscillatory pattern, while the other two curves show uniform oscillations. In conclusion, the study of incommensurate order models provides a more accurate replication of the natural complexity observed in biological systems. These models uncover hidden characteristics of the system's state variables and capture the intricate dynamics with greater precision. By considering non-identical fractional orders, we can better understand the complex behavior and unique properties exhibited by the biological populations.

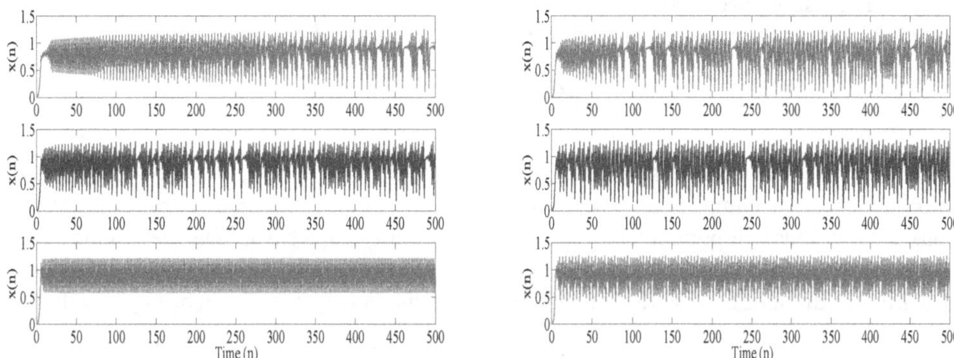

Fig. 5.11: Time series plots corresponding to bifurcation diagrams of the system (5.12) in Figure 5.9 for $\theta = 2.35, 2.5$.

5.5 Tumor Immune Model with Chemotherapy

Various treatment methods are available for combating cancer cells, including radiation therapy, immunotherapy, chemotherapy, and surgery, among others. Each treatment approach has its limitations. For example, radiation therapy can negatively affect nearer healthy cells, while surgery carries the risk of spreading tumor cells to adjacent areas. Among these methods, chemotherapy is commonly used when there is rapid cancer cell growth or a risk of cancer recurrence. Chemotherapy involves the administration of powerful drugs orally or intravenously to kill cancer cells.

Chemotherapeutic treatments have proven to be effective against different types of cancers. However, one major concern with chemotherapy is the potential side effects associated with the drugs used. Some side effects are immediate and can be managed, while others may occur later and have more complex effects. Chemotherapy is employed for various reasons, such as eliminating hidden cancer cells after other treatments, reducing tumor size before surgery, relieving symptoms and pain caused by tumors, and even treating other conditions. The administration of chemotherapeutic drugs can take different forms, including tablets, intravenous infusions, and topical applications (such as ointments for certain skin cancers).

Determining the appropriate dosage of chemotherapy drugs is crucial, as they can affect both tumor cells and normal healthy cells. Overdosing can lead to complete destruction of tumor cells but may also harm healthy cells, resulting in side effects. Conversely, underdosing may have no impact on tumor cells, rendering the treatment ineffective. Mathematical models have been developed to understand the dynamics of the tumor-immune system in the presence of chemotherapy. Specifically, this section focuses on incommensurate order models.

Several notable research contributions have investigated the modeling of chemotherapy effects on tumor-immune dynamics. Studies such as [8] and [10] examined the optimal response of chemotherapy and explored optimal control aspects. The work presented in [12] discussed different treatments for cancer cells, while [13] focused on anti-cancer treatments in general. Fractional order models replicating tumor-immune systems have gained increasing interest, as evidenced by works such as [25] and [26]. Additionally, research has explored fuzzy sets-based fractional order dynamics in [24], and numerical investigations on chemotherapeutic models with fractional derivatives were conducted in [27] and [29].

Chemotherapeutic Drug Concentration:
The mathematical representation of the chemotherapeutic drugs for cancer treatment can be represented with a simple differential equation of the form,

$$\frac{dC}{dn} = \theta_1 - \xi C. \tag{5.19}$$

The concentration of the drugs administered at time (n) is denoted by C, while θ_1 represents the dosage of the drugs introduced into the human body. The parameter ξ signifies the decay rate of drug concentration due to interactions with various cell populations. The objective of this chapter is to analyze a tumor kill model that leads to a reduction in tumor cell populations over time. The proposed model is a modified discrete-time fractional order tumor-immune cell interaction model with chemotherapy, which can be expressed as follows:

$$\begin{cases} \Delta^{\vartheta_1} x(n) = & \theta x(n-1+\vartheta_1)(1-x(n-1+\vartheta_1)) - bx(n-1+\vartheta_1)\Psi(x(n-1+\vartheta_1), y(n-1+\vartheta_1)) \\ & -cx(n-1+\vartheta_1)y(n-1+\vartheta_1) - \Phi_1(C)x(n-1+\vartheta_1), \\ \Delta^{\vartheta_2} y(n) = & -\Phi_2(C)y(n-1+\vartheta_2) - gx(n-1+\vartheta_2) - wx(n-1+\vartheta_2)y(n-1+\vartheta_2) \\ & +u\dfrac{(\Psi(x(n-1+\vartheta_2), y(n-1+\vartheta_2)))^2 x(n-1+\vartheta_2)^2}{h+(\Psi(x(n-1+\vartheta_2), y(n-1+\vartheta_2)))^2 x(n-1+\vartheta_2)^2} y(n-1+\vartheta_2), \\ \Delta^{\vartheta_3} C(n) = & \theta_1 - \xi C(n-1+\vartheta), \end{cases} \tag{5.20}$$

where $\Phi_1(C)$ and $\Phi_2(C)$ are the kill rates of the tumor and effector cells due to drug introduction. In this chapter, the kill rate for effector cells and tumor cells due to the concentration of the drugs is considered in two different scenarios.

1. Constant kill rate : $\Phi_i(C) = \xi_i C, i = 1, 2,$
2. Exponential kill rate : Tumor: $\Phi_1(C) = \xi_1\left(1 - e^{\xi_2 C}\right),$
   Effector cells: $\Phi_2(C) = \xi_3\left(1 - e^{\xi_4 C}\right).$

To assess the impact of drug input on the sustainability of tumor and effector cell populations, we aim to examine the behavior of the system using bifurcation diagrams. In the context of chemotherapy, determining the appropriate dosage is crucial to achieve an effective treatment with minimal side effects. Therefore, we investigate the dynamics of the system by varying the input rate parameter θ_1 for the two aforementioned scenarios.

5.5.1 Chemotherapy Treatment with Constant Kill Rate

In this section, we consider the concentration of the drug to have a constant kill rate on both tumor and effector cells over time. This assumption is commonly used in mathematical models to depict the effect of introduced drugs on tumor cells. The mathematical model representing the introduction of chemotherapy treatment with a constant kill rate takes the following form:

$$\begin{cases} \Delta^{\vartheta_1} x(n) = & \theta x(n-1+\vartheta_1)(1-x(n-1+\vartheta_1)) - bx(n-1+\vartheta_1)\Psi(x(n-1+\vartheta_1), y(n-1+\vartheta_1)) \\ & -cx(n-1+\vartheta_1)y(n-1+\vartheta_1) - \xi_1 x(n-1+\vartheta_1)C(n-1+\vartheta_1), \\ \Delta^{\vartheta_2} y(n) = & -\xi_2 y(n-1+\vartheta_2)C(n-1+\vartheta_2) - gx(n-1+\vartheta_2) - wx(n-1+\vartheta_2)y(n-1+\vartheta_2) \\ & +u\dfrac{(\Psi(x(n-1+\vartheta_2), y(n-1+\vartheta_2)))^2 x(n-1+\vartheta_2)^2}{h+(\Psi(x(n-1+\vartheta_2), y(n-1+\vartheta_2)))^2 x(n-1+\vartheta_2)^2} y(n-1+\vartheta_2), \\ \Delta^{\vartheta_3} C(n) = & \theta_1 - \xi C(n-1+\vartheta_3), \end{cases} \tag{5.21}$$

To facilitate the mathematical discussion and numerical simulations, a more computationally suitable form of the system is obtained from the proposed equation (5.21). This modified form of the system allows for easier analysis and simulation, and it is given by:

$$\begin{cases} x(n) = & x(0) + \dfrac{1}{\Gamma(\vartheta_1)} \sum_{k=1}^{n} \dfrac{\Gamma(n-k+\vartheta_1)}{\Gamma(n-k+1)} \Big[\theta x(k-1)(1-x(k-1)) - bx(k-1)\Psi(x(k-1), y(k-1)) \\ & -cx(k-1)y(k-1) - \xi_1 x(k-1)C(k-1) \Big], \\ y(n) = & y(0) + \dfrac{1}{\Gamma(\vartheta_2)} \sum_{k=1}^{n} \dfrac{\Gamma(n-k+\vartheta_2)}{\Gamma(n-k+1)} \Big[-gx(k-1) - wx(k-1)y(k-1) \\ & +u \dfrac{(\Psi(x(k-1), y(k-1)))^2 x(k-1)^2}{h + (\Psi(x(k-1), y(k-1)))^2 x(k-1)^2} y(k-1) - \xi_2 y(k-1)C(k-1) \Big], \\ C(n) = & C(0) + \dfrac{1}{\Gamma(\vartheta_3)} \sum_{k=1}^{n} \dfrac{\Gamma(n-k+\vartheta_3)}{\Gamma(n-k+1)} \Big[\theta_1 - \xi C(k-1) \Big]. \end{cases} \tag{5.22}$$

where $n = 1, 2, 3 \cdots$.

The chaotic behavior of the system is analyzed with respect to the input rate of the chemotherapeutic drug, denoted as θ_1, which varies within the range of $[0.36, 2]$. The remaining parameters of the system are set as follows: $b = 0.13889$, $c = 0.34$, $d = 0.36$, $g = 0.5$, $w = 0.475$, $u = 0.02$, $h = 0.72$, $p = 0.5$, $s = 0.36$, $\theta = 2.9$, $\xi_1 = 0.25$, $\xi_2 = 0.01$, $\xi = 0.3$, $\vartheta_1 = 0.7$, $\vartheta_2 = 0.1$, and $\vartheta_3 = 0.5$. The dynamics of the tumor cell population and effector cell populations, as influenced by the input rate of the drug, are visualized through the bifurcation diagrams presented in Figure 5.12. It can be observed that for small doses of drugs, the system exhibits chaotic behavior characterized by unpredictable fluctuations in the populations of both tumor cells and effector cells.

As the input rate of the drugs increases, the tumor and effector cells exhibit a stable behavior, resulting in constant cell populations. This indicates that with sufficient drug dosage, the tumor cells can be effectively treated and completely eliminated from the individual. The bifurcation diagrams provide an overview of the system's behavior, but to further analyze the system, the analysis is performed by breaking down the results for different input values of θ_1 over time. This is depicted through time series plots in Figures 5.13, 5.14, and 5.15.

The system's behavior undergoes a transformation from unpredictable random motion at $\theta_1 = 0.45$ to gradually transitioning into period-4 oscillations at $\theta_1 = 0.85$, and finally reaching a stable nature at $\theta_1 = 1.85$.

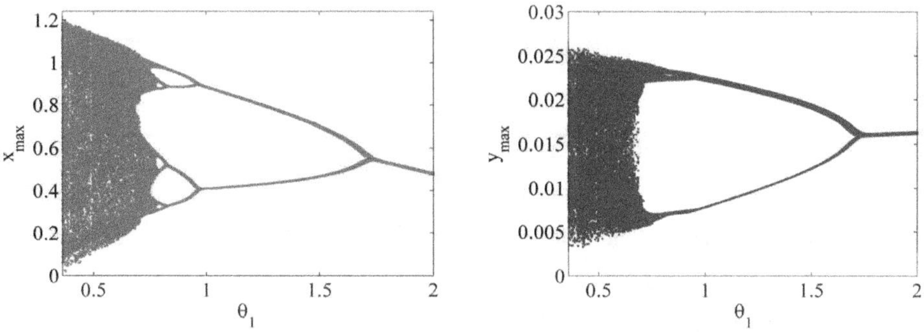

Fig. 5.12: Bifurcation diagrams of the system (5.21).

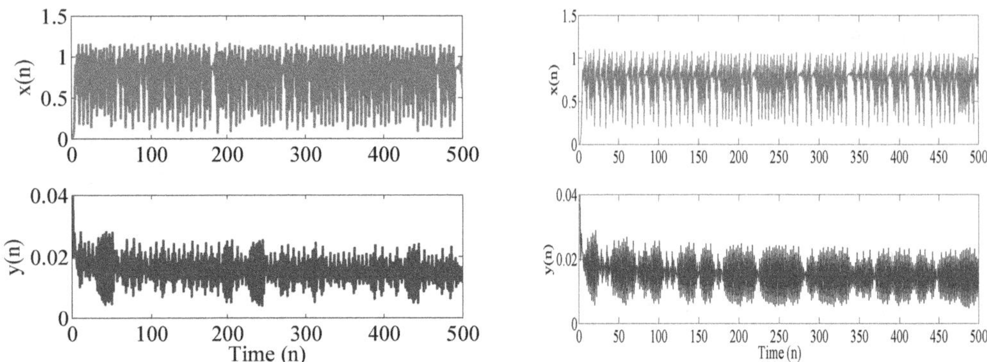

Fig. 5.13: Time series plots corresponding to bifurcation diagrams of the system (5.21) in Figure 5.12 for $\theta_1 = 0.45, 0.65$.

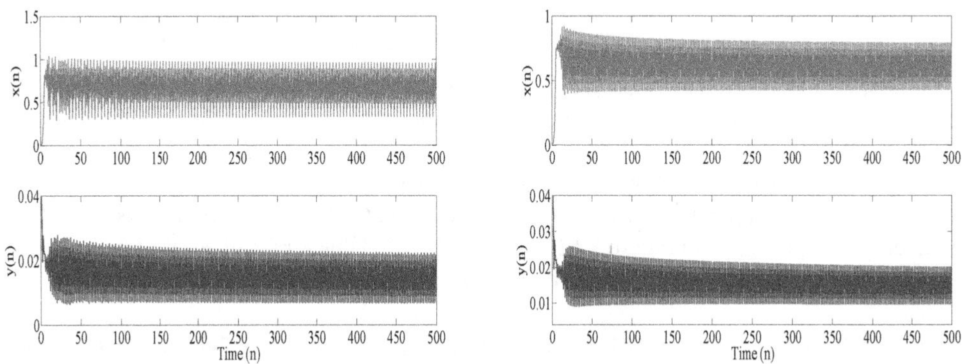

Fig. 5.14: Time series plots corresponding to bifurcation diagrams of the system (5.21) in Figure 5.12 for $\theta_1 = 0.85, 1.25$.

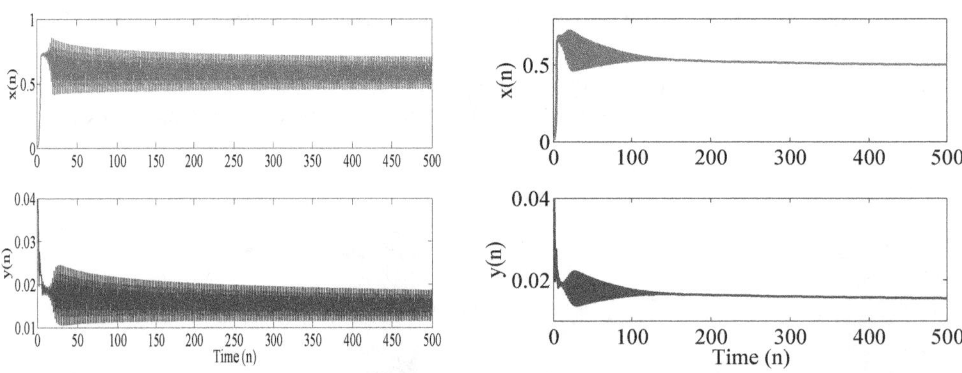

Fig. 5.15: Time series plots corresponding to bifurcation diagrams of the system (5.21) in Figure 5.12 for $\theta_1 = 1.45, 1.85$.

These plots provide a detailed insight into the dynamics of the system as the input rate of the drugs varies, showcasing the progression from chaotic behavior to stable patterns with increasing drug dosage.

5.5.2 Chemotherapy Treatment with Exponential Kill Rate

The primary motivation behind studying tumor growth models is to develop effective therapies for controlling their growth. A crucial aspect of modeling is understanding the effectiveness of drugs in reducing the growth of tumor cells while considering their potential toxicity. To explore these characteristics, it is essential to develop models that accurately represent the realistic behavior of cells and the impact of drugs. This section of the chapter focuses on investigating the introduction of drugs with an exponential kill rate on tumor and effector cells. The modified model is given by

$$\begin{cases} \Delta^{\vartheta_1}x(n) = & \theta x(n-1+\vartheta_1)(1-x(n-1+\vartheta_1)) - bx(n-1+\vartheta_1)\Psi(x(n-1+\vartheta_1), y(n-1+\vartheta_1)) \\ & -cx(n-1+\vartheta_1)y(n-1+\vartheta_1) - \xi_1 x(n-1+\vartheta_1)\left(1-e^{\xi_2 C(n-1+\vartheta_1)}\right), \\ \Delta^{\vartheta_2}y(n) = & -\xi_3 y(n-1+\vartheta_2)\left(1-e^{\xi_4 C(n-1+\vartheta_2)}\right) - gx(n-1+\vartheta_2) - wx(n-1+\vartheta_2)y(n-1+\vartheta_2) \\ & +u\dfrac{(\Psi(x(n-1+\vartheta_2), y(n-1+\vartheta_2)))^2 x(n-1+\vartheta_2)^2}{h+(\Psi(x(n-1+\vartheta_2), y(n-1+\vartheta_2)))^2 x(n-1+\vartheta_2)^2} y(n-1+\vartheta_2), \\ \Delta^{\vartheta_3}C(n) = & \theta_1 - \xi C(n-1+\vartheta_3), \end{cases} \tag{5.23}$$

where ξ_1 and ξ_3 represent the rate at which the tumor and effector cells are killed, and ξ_2 and ξ_4 are the rates at which the tumor and effector cell populations are capable of resisting the drug impact. Respective numerical forms for analysis are given by

$$\begin{cases} x(n) = & x(0) + \dfrac{1}{\Gamma(\vartheta_1)}\sum_{k=1}^{n}\dfrac{\Gamma(n-k+\vartheta_1)}{\Gamma(n-k+1)}\Big[\theta x(k-1)(1-x(k-1)) - bx(k-1)\Psi(x(k-1), y(k-1)) \\ & -cx(k-1)y(k-1) - \xi_1 x(k-1)\left(1-e^{\xi_2 C(k-1)}\right)\Big], \\ y(n) = & y(0) + \dfrac{1}{\Gamma(\vartheta_2)}\sum_{k=1}^{n}\dfrac{\Gamma(n-k+\vartheta_2)}{\Gamma(n-k+1)}\Big[-\xi_3 y(k-1)\left(1-e^{\xi_4 C(k-1)}\right) - gx(k-1) - wx(k-1)y(k-1) \\ & +u\dfrac{(\Psi(x(k-1), y(k-1)))^2 x(k-1)^2}{h+(\Psi(x(k-1), y(k-1)))^2 x(k-1)^2} y(k-1)\Big], \\ C(n) = & C(0) + \dfrac{1}{\Gamma(\vartheta_3)}\sum_{k=1}^{n}\dfrac{\Gamma(n-k+\vartheta_3)}{\Gamma(n-k+1)}\Big[\theta_1 - \xi C(k-1)\Big]. \end{cases} \tag{5.24}$$

where $n = 1, 2, 3\cdots$.

We will investigate the chaotic behavior of the system by analyzing the impact of the input rate of the chemotherapeutic drug, denoted as θ_1, within the range of [0, 6]. The system parameters are set as follows: $b = 0.13889$, $c = 0.34$, $d = 0.36$, $g = 0.5$, $w = 0.475$, $u = 0.02$, $h = 0.72$, $p = 0.5$, $s = 0.36$, $\theta = 2.4$, $\xi_1 = 0.875$, $\xi_2 = 0.04$, $\xi_3 = 0.04$, $\xi_4 = 0.02$, $\xi = 0.15$ and $\vartheta_1 = 0.7$, $\vartheta_2 = 0.1$, $\vartheta_3 = 0.5$. The bifurcation diagrams, depicted in Figure 5.16, illustrate the behavior of the tumor cells with these parameter values. It can be observed that the tumor cells exhibit aperiodic behavior when the input rate of drugs is low.

The impact of the input rate of the chemotherapeutic drug on the effector cell populations is relatively minimal compared to the tumor cell populations. The bifurcation diagrams in Figures 5.17, 5.18, 5.19 illustrate the time-varying behavior of the effector cells under different input rates. It can be observed that the regions of chaos for the effector cells are quite small. In particular, for a small value of $\theta_1 = 0.03$, the effector cells exhibit random behavior. However, as the value of θ_1 gradually increases, the behavior of the effector cells stabilizes and tends towards extinction. On the other hand, the tumor cells exhibit chaotic behavior when

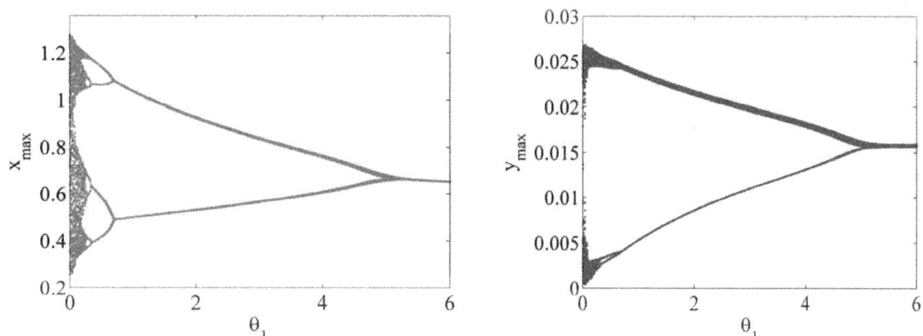

Fig. 5.16: Bifurcation diagrams of the system (5.23).

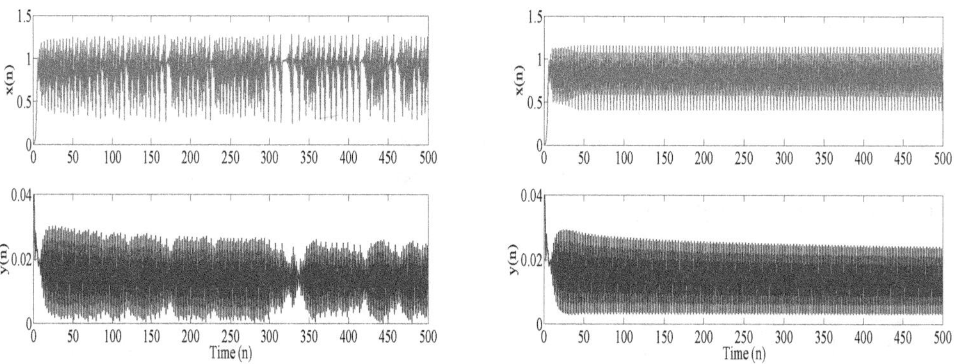

Fig. 5.17: Time series plots corresponding to bifurcation diagrams of the system (5.23) in Figure 5.16 for $\theta_1 = 0.03, 0.5$.

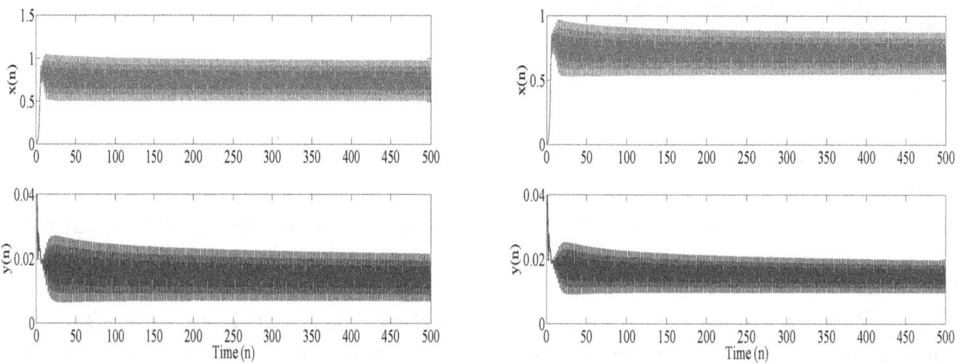

Fig. 5.18: Time series plots corresponding to bifurcation diagrams of the system (5.23) in Figure 5.16for $\theta_1 = 1.5, 2.5$.

subjected to exponential kill, transitioning from random chaotic behavior to uniform oscillations and finally reaching a stable state.

The introduction of drugs with an exponential decay effect on tumor cells results in the control of tumor cell populations. However, this control comes at the expense of the extinction of effector cells. This

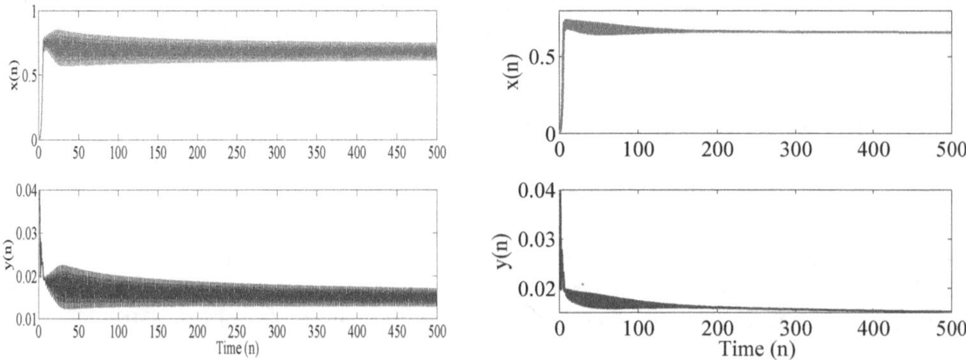

Fig. 5.19: Time series plots corresponding to bifurcation diagrams of the system (5.23) in Figure 5.16 for $\theta_1 = 4, 5.5$.

observation highlights the importance of completely eliminating the tumor cells when their behavior becomes predictable. Failure to do so may lead to the recurrence of cancer cells, which can have severe and potentially fatal consequences.

5.6 Summary

The study focused on the intriguing and captivating topic of tumor growth in biological systems. By introducing complexity to the 2-D system, the fractional kill rate of tumor cells and the activation of effector cells, along with the competition factor among the cell populations, were incorporated. The investigation considered both commensurate and incommensurate orders, providing a better understanding of the characteristics of the state variables. The system displayed complex dynamics, which were analyzed using bifurcation diagrams. The choice of fractional orders was found to significantly influence the system's behavior, as demonstrated by numerical simulations and a sequence of bifurcation diagrams for different fractional orders. Furthermore, the impact of the chemotherapy drug concentration was explored. It was observed that lower drug concentrations resulted in chaotic behavior, while higher concentrations had different effects depending on the kill rate. For a constant kill rate, higher drug concentrations led to stable sustainability of both populations. However, when an exponential kill law was considered, it resulted in the extinction of effector cells, allowing the tumor population to survive at a constant rate. Overall, the study made interesting contributions to the modeling of tumor and immune cell interactions under various scenarios.

References

[1] Jan Cermak, Istvan Gyori and Ludek Nechvatal. On explicit stability conditions for a linear fractional difference system. Fractional Calculus and Applied Analysis 18 (2015): 651–672.

[2] Adel Ouannas, Amina Aicha Khennaoui, Shaher Momani, Giuseppe Grassi and Viet-Thanh Pham. Chaos and control of a three-dimensional fractional order discrete-time system with no equilibrium and its synchronization. AIP Advances 10, no. 4 (2020): 045310.

[3] Denise Kirschner and John Carl Panetta. Modeling immunotherapy of the tumor–immune interaction. Journal of Mathematical Biology 37, no. 3 (1998): 235–252.

[4] Lisette G. de Pillis, Ami E. Radunskaya and Charles L. Wiseman. A validated mathematical model of cell-mediated immune response to tumor growth. Cancer Research 65, no. 17 (2005): 7950–7958.

[5] Mohd Taib Shatnawi, Noureddine Djenina, Adel Ouannas, Iqbal M. Batiha and Giuseppe Grassi. Novel convenient conditions for the stability of nonlinear incommensurate fractional-order difference systems. Alexandria Engineering Journal 61, no. 2 (2022): 1655–1663.

[6] Mark Robertson-Tessi, Ardith El-Kareh and Alain Goriely. A mathematical model of tumor–immune interactions. Journal of Theoretical Biology 294 (2012): 56–73.

[7] Grace E. Mahlbacher, Kara C. Reihmer and Hermann B. Frieboes. Mathematical modeling of tumor-immune cell interactions. Journal of Theoretical Biology 469 (2019): 47–60.

[8] Urszula Ledzewicz, Mohammad Naghnaeian and Heinz Schattler. Optimal response to chemotherapy for a mathematical model of tumor–immune dynamics. Journal of Mathematical Biology 64, no. 3 (2012): 557–577.

[9] Amina Eladdadi, Peter Kim and Dann Mallet (eds.). Mathematical models of tumor-immune system dynamics. Vol. 107. New York: Springer (2014).

[10] Urszula Ledzewicz, Mozhdeh Sadat Faraji Mosalman and Heinz Schattler. Optimal controls for a mathematical model of tumor-immune interactions under targeted chemotherapy with immune boost. Discrete & Continuous Dynamical Systems-B 18, no. 4 (2013): 1031.

[11] Lisette G. de Pillis and Ami E. Radunskaya. Modeling tumor-immune dynamics. In Mathematical Models of Tumor-Immune System Dynamics, pp. 59–108. Springer New York (2014).

[12] Isaeva, O. G. and V. A. Osipov. Different strategies for cancer treatment: mathematical modelling. Computational and Mathematical Methods in Medicine 10, no. 4 (2009): 253–272.

[13] Urszula Ledzewicz and Heinz Schaettler. Optimizing chemotherapeutic anti-cancer treatment and the tumor microenvironment: An analysis of mathematical models. Systems Biology of Tumor Microenvironment (2016): 209–223.

[14] Ahmed, E., A. Hashish and F. A. Rihan. On fractional order cancer model. Journal of Fractional Calculus and Applied Analysis 3, no. 2 (2012): 1–6.

[15] Rihan, F. A. and G. Velmurugan. Dynamics of fractional-order delay differential model for tumor-immune system. Chaos, Solitons & Fractals 132 (2020): 109592.

[16] Ilhan Ozturk and Fatma Ozkose. Stability analysis of fractional order mathematical model of tumor-immune system interaction. Chaos, Solitons & Fractals 133 (2020): 109614.

[17] Esmehan Ucar, Necati Ozdemir and Eren Altun. Fractional order model of immune cells influenced by cancer cells. Mathematical Modelling of Natural Phenomena 14, no. 3 (2019): 308.

[18] Fathalla A. Rihan. Numerical modeling of fractional-order biological systems. In Abstract and Applied Analysis, vol. 2013. Hindawi (2013).

[19] Fathalla A. Rihan, Adel Hashish, Fatma Al-Maskari, M. S. Hussein, Elsayed Ahmed, M. B. Riaz and Radoune Yafia. Dynamics of tumor-immune system with fractional-order. Journal of Tumor Research 2, no. 1 (2016): 109–115.

[20] Ercan Balci, Ilhan Ozturk and Senol Kartal. Dynamical behaviour of fractional order tumor model with Caputo and conformable fractional derivative. Chaos, Solitons & Fractals 123 (2019): 43–51.

[21] Muhammad Arfan, Kamal Shah, Aman Ullah, Meshal Shutaywi, Poom Kumam and Zahir Shah. On fractional order model of tumor dynamics with drug interventions under nonlocal fractional derivative. Results in Physics 21 (2021): 103783.

[22] Jehad Alzabut, A. Selvam, Vignesh Dhakshinamoorthy, Hakimeh Mohammadi and Shahram Rezapour. On chaos of discrete time fractional order host-immune-tumor cells interaction model. Journal of Applied Mathematics and Computing (2022): 1–26.

[23] Ercan Balci, Senol Kartal and Ilhan Ozturk. Comparison of dynamical behavior between fractional order delayed and discrete conformable fractional order tumor-immune system. Mathematical Modelling of Natural Phenomena 16 (2021): 3.
[24] Dhanalakshmi, P., S. Senpagam and R. Mohanapriya. Finite-time fuzzy reliable controller design for fractional-order tumor system under chemotherapy. Fuzzy Sets and Systems 432 (2022): 168–181.
[25] Fatma Ozkose, Mehmet Tamer Senel and Rafla Habbireeh. Fractional-order mathematical modelling of cancer cells-cancer stem cells-immune system interaction with chemotherapy. Mathematical Modelling and Numerical Simulation with Applications 1, no. 2 (2021): 67–83.
[26] Changjin Xu, Muhammad Farman, Ali Akgul, Kottakkaran Sooppy Nisar and Aqeel Ahmad. Modeling and analysis fractal order cancer model with effects of chemotherapy. Chaos, Solitons & Fractals 161 (2022): 112325.
[27] Zulqurnain Sabir, Maham Munawar, Mohamed A. Abdelkawy, Muhammad Asif Zahoor Raja, Canan Unlu, Mdi Begum Jeelani and Abeer S. Alnahdi. Numerical investigations of the fractional-order mathematical model underlying immune-chemotherapeutic treatment for breast cancer using the neural networks. Fractal and Fractional 6, no. 4 (2022): 184.
[28] Changjin Xu, Peiluan Li, Maoxin Liao and Shuai Yuan. Bifurcation analysis for a fractional-order chemotherapy model with two different delays. Mathematical Methods in the Applied Sciences 43, no. 3 (2020): 1053–1083.
[29] Sachin Kumar and Abdon Atangana. A numerical study of the nonlinear fractional mathematical model of tumor cells in presence of chemotherapeutic treatment. International Journal of Biomathematics 13, no. 03 (2020): 2050021.

Chapter 6

Attractors in Fractional Order Discrete-Time Rabinovich-Fabrikant System

This chapter focuses on the introduction of discrete–time fractional order complex state variables to the Rabinovich-Fabrikant system and the examination of their dynamics. The Rabinovich-Fabrikant system is widely applicable in various fields such as hydrodynamic flow, diffusion reactions, and chemical reactions. By incorporating a Caputo type discrete fractional operator, the system's stability is analyzed under two cases: one with a unique fractional order and another with distinct fractional orders. The differences between these cases are explored through numerical simulations, including time series plots and 3-Dimensional attractors. A significant aspect of this study involves comprehending the chaotic behavior, which is supported by bifurcation diagrams showcasing the system's fractional orders. Additionally, the impact of parameters on the system dynamics is investigated, particularly in the case of the complex form of the Rabinovich-Fabrikant system. Furthermore, the sensitivity of the system to initial conditions is illustrated by the presence of coexisting attractors.

6.1 Nonlinearity in Nature

Throughout history, human beings have demonstrated a strong interest and curiosity in understanding the natural events that occur in the real world. The progress of science and technology is driven by our desire to explore and comprehend these phenomena, which are often unpredictable and difficult to explain. Researchers from various interdisciplinary fields conduct a wide range of experiments in an attempt to recreate these events, but not all experiments yield high success rates. This low success rate leads to wasted time and financial resources. This is where the role of mathematics becomes crucial in the construction of realistic models, enabling researchers to investigate the dynamics of the phenomena under consideration. Mathematical models allow for the inclusion of different parameters that influence the behavior of the natural phenomena, making them a cost-efficient tool. The development of supercomputers in the 1950s further enhanced the ability to perform highly accurate simulations, providing interdisciplinary scientists with valuable insights on their experimental fronts.

It is evident that not all real-world events exhibit linear behavior. This is apparent when considering phenomena such as climate change, which demonstrates random or unpredictable behavior that is challenging to predict. Another example is the coexistence of species in an ecosystem, which is a nonlinear phenomenon. Therefore, it can be concluded that almost all natural phenomena are nonlinear in nature.

6.1.1 Chaos and Nature

The study of random and unpredictable systems has given rise to the interdisciplinary field of chaos theory. Chaos can be observed in various everyday events such as heartbeats, fluid flow, road traffic, stock markets, and ecology. This field has captivated the attention of scientists, engineers, and researchers due to its ability to reveal unknown features of the physical systems being studied. From a mathematical perspective, chaos theory describes the outcomes of dynamical systems resulting from small changes in their initial conditions. One common approach to studying the chaotic dynamics of a system is through bifurcation analysis. Bifurcation can be understood as the visualization of qualitative changes in a system, transitioning from a state of stability to periodic orbits or aperiodic states, or vice versa, as specific parameters vary. This type of analysis provides valuable insights into how the behavior of a system is influenced by the variation of a single parameter.

Dynamical systems, often represented by systems of ordinary differential equations, are commonly used to model real-life events or phenomena. Numerous chaotic systems have been proposed and investigated, such as the Lorentz system, Rossler's system, Liu system, Chua system, and Rabinovich-Fabrikant system. Technological advancements since the mid-20th century have led to the development of fractional order derivatives. While the concept of fractional order derivatives is not new, it has been revitalized with the support of theoretical and geometrical interpretations made possible by computers. Fractional derivatives introduce an additional parameter for study, which proves helpful in capturing the system's past memory for future predictions. Fractional order derivatives also allow for the expression of physical properties that ordinary differential equations fail to elucidate. On the experimental front, results obtained using fractional order models demonstrate better agreement with observed phenomena compared to models using traditional derivatives. Additionally, the construction of fractional order models in discrete–time has significantly contributed to the exploration of system dynamics.

6.1.2 Dynamical Systems with Complex Variables

Real numbers are universally used in most mathematical theories, while complex numbers come into play when analyzing certain theoretical concepts. Modeling theories often rely on real-valued state variables, with complex-valued state variables being rarely proposed. However, the ever-growing curiosity of humanity has led to the construction of models with complex state variables. In such cases, real and imaginary parts of the complex state variables are separated for further investigation. Chaotic systems with complex state variables exhibit different dynamical behaviors compared to systems with real state variables. Complex-variable dynamical systems find applications in various fields such as disk dynamos, lasers, rotating fluids [1, 2], information encryption, and secure communications [3–5].

Some examples of complex chaotic systems include the complex Lorenz equation [6], complex Chen system, Lu system, and hyperchaotic complex Chen and Lu system [7–10]. However, the exploration of fractional order complex chaotic systems has not been as actively pursued and solutions with memory effects of impulsive fractional differential equations was illustrated in [25].

The Rabinovich-Fabrikant system has been extensively studied by several researchers, with different approaches and applications. Graphical structures of the basins of attraction have been discussed by M.F. Danca in [11]. Hidden attractors and complex dynamics exhibited by the system have been explored in [12–15]. Srivastava et al [16] analysed the fractional order version of the equation to perform chaos control and synchronisation, Moaddy et al [17] performed the numerical investigation of the fractional order equation, and [18] explains the synchronisation results for the fractional order system. In [19], the application of the system to image encryption was analysed. The system has also been applied to image encryption. Recently, dynamics of variable order discrete–time systems have been investigated in [24, 26, 27] and application of discrete fractional calculus to neural networks employing the fixed point technique was discussed in [28]. However, no research has yet considered the complex form of the Rabinovich-Fabrikant system. To further expand the modeling with complex state variables, this chapter aims to study and illustrate the dynamics of a nonlinear Rabinovich-Fabrikant system with a discrete fractional order operator, which is subsequently transformed into a system with complex state variables.

6.2 Rabinovich–Fabrikant System

The Rabinovich-Fabrikant system is named after scientists Mikhail Rabinovich and Anatoly Fabrikant, who introduced the equations in 1979. These equations have found applications in various fields, including hydrodynamic flows and chemical reactions involving diffusion. The system consists of three first-order differential equations with two parameters and exhibits complex behavior due to the presence of cubic and quadratic nonlinearities. This complexity poses a significant challenge for researchers to fully analyze and understand its characteristics. To address this, the mathematical system proposed by Rabinovich and Fabrikant is investigated using the discrete–time Caputo difference operator, which takes the following form:

$$\begin{cases} \Delta^{\vartheta}x(n) = & y(n-1+\vartheta)\left(z(n-1+\vartheta)-1+(x(n-1+\vartheta))^2\right)+\delta\, x(n-1+\vartheta), \\ \Delta^{\vartheta}y(n) = & x(n-1+\vartheta)\left(3z(n-1+\vartheta)+1-(x(n-1+\vartheta))^2\right)+\delta y(n-1+\vartheta), \\ \Delta^{\vartheta}z(n) = & -2z(n-1+\vartheta)(p+x(n-1+\vartheta)y(n-1+\vartheta)), \end{cases} \tag{6.1}$$

where Δ^{ϑ} is a Caputo difference operator with $0 < \vartheta \leq 1$, $n = 0, 1, 2, \cdots$, $\delta, p > 0$ are real parameters, and $n \in \mathbb{N}_{1-\vartheta}$. The proposed form of the equation in (6.1) cannot be numerically analysed, so we use the fractional sum equation method to convert the equation to a simplified numerical form suitable for numerical simulations [22]. The numerical form is derived as follows:

$$\begin{cases} x(n) = & x(0) + \dfrac{1}{\Gamma(\vartheta)}\sum\limits_{k=1}^{n}\dfrac{\Gamma(n-k+\vartheta)}{\Gamma(n-k+1)}\Big(y(j-1)\left(z(j-1)-1+(x(j-1))^2\right) \\ & +\delta\, x(j-1)\Big), \\ y(n) = & y(0) + \dfrac{1}{\Gamma(\vartheta)}\sum\limits_{k=1}^{n}\dfrac{\Gamma(n-k+\vartheta)}{\Gamma(n-k+1)}\Big(x(j-1)\left(3z(j-1)+1-(x(j-1))^2\right) \\ & +\delta y(j-1)\Big), \\ z(n) = & z(0) + \dfrac{1}{\Gamma(\vartheta)}\sum\limits_{k=1}^{n}\dfrac{\Gamma(n-k+\vartheta)}{\Gamma(n-k+1)}\left(-2z(j-1)(p+x(j-1)y(j-1))\right), \end{cases} \tag{6.2}$$

where $n = 1, 2, \cdots$. The system represented by equations (6.1), where the equations have identical fractional orders, is referred to as a commensurate-type fractional order discrete–time system. On the other hand, when the equations have distinct fractional orders, it is classified as an incommensurate order fractional order system. The system displays distinct dynamical behaviors depending on how the fractional orders are considered for the state variables. To gain insights into the system's behavior, qualitative analysis is performed by studying the equilibrium points, which are then numerically validated under certain assumptions of parameter values.

6.3 Equilibrium Points and Stability

The stability theory plays a crucial role in analyzing real-world phenomena, as it provides essential qualitative properties. In this section, we focus on numerically evaluating the equilibrium points and their stability in both the commensurate order and incommensurate fractional order discrete–time Rabinovich-Fabrikant system. The system represented by equations (6.1) possesses equilibrium points, including the trivial equilibrium point at $Q_0 = (0, 0, 0)$ and interior equilibrium points as follows:

$$Q_{1,2} = \left(\mp\sqrt{\frac{p(\beta+2)}{4p-3\delta}}, \pm\sqrt{p\frac{4p-3\delta}{\beta+2}}, \frac{a\beta+\beta_1}{(4p-3\delta)\beta+8p-6\delta} \right),$$

$$Q_{3,4} = \left(\mp\sqrt{\frac{p(\beta-2)}{3\delta-4p}}, \pm\sqrt{p\frac{4p-3\delta}{2-\beta}}, \frac{a\beta-\beta_1}{(4p-3\delta)\beta-8p+6\delta} \right).$$

where $\beta = \sqrt{3\delta^2 - 4\delta p + 4}$ and $\beta_1 = 4\delta p^2 - 7\delta^2 p + 3\delta^3 + 2\delta$. The presented equilibrium points may not always be real numbers due to the presence of square roots and the parameters δ and p, which are real numbers that can take on negative values. Throughout the numerical verification of the results in this chapter, we will concentrate on equilibrium point analysis for real values.

6.3.1 Commensurate Order System (6.1)

To investigate the stability analysis of the system (6.1), consider the parameter values $\delta = -0.6$ and $p = 0.3$ with fractional order $\vartheta = 0.6$. $Q_0 = (0, 0, 0)$ and $Q_{1,2} = (\pm 0.66395, \mp 0.45184, -0.32249)$ are the obtained real valued interior equilibriums. We disregard the system's complex equilibrium points when performing mathematical validation of stability results. To numerically investigate the stability of the system (6.1) for the considered parameter values, the following theorem is required.

Theorem 6.1 *[21]Let $\vartheta \in (0, 1)$ and $\Psi \in \mathbb{R}^{k \times k}$. Let*

$$B^{\vartheta} = \left\{ \lambda \in \mathbb{C} : |\lambda| < \left(2\cos\left(\frac{|arg\lambda| - \phi}{2-\vartheta} \right) \right)^{\vartheta} \text{ and } |arg\lambda| > \frac{\vartheta\pi}{2} \right\}.$$

If all the eigenvalues of Ψ lie in B^{ϑ}, then the system (6.1) *is asymptotically stable. The system* (6.1) *is unstable if $\lambda \in \mathbb{C} \backslash cl(B^{\vartheta})$ for an eigenvalue of Ψ.*

Numerical validation for different equilibrium points are carried out separately employing the Theorem 6.1. **Trivial Equilibrium Point** (Q_0)**:** To investigate the stability results for the considered system, we shall obtain eigenvalues from the Jacobian matrix obtained at the trivial equilibrium point,

$$J(Q_0) = \begin{bmatrix} -0.6 & -1 & 0 \\ 1 & -0.6 & 0 \\ 0 & 0 & -0.6 \end{bmatrix}, \tag{6.3}$$

yielding the eigenvalues $\eta_{1,2} = -0.6 \pm i$ and $\eta_3 = -0.6$. From Theorem 6.1, the presence of the eigenvalues in the set B^{ϑ} needs to be verified. Absolute values of the argument of the eigenvalues are $|arg(-0.6 \pm i)| = 2.11121$ and $|arg(-0.6)| = 3.14159$, which are clearly greater than $\frac{0.6\pi}{2}$. For the other condition, $\left(2\cos\left(\frac{|arg\eta_{1,2}|-\phi}{2-0.6}\right)\right)^{0.6} = 1.26577 > |-0.6 \pm i| = 1.16619$ and $\left(2\cos\left(\frac{|arg\eta_3|-\phi}{2-0.6}\right)\right)^{0.6} = 1.51571 > |-0.6|$. Hence, it is evident that the trivial equilibrium point Q_0 is asymptotically stable.

Interior Equilibrium Point ($Q_{1,2}$): The Jacobian matrix at Q_1 is

$$J(Q_1) = \begin{bmatrix} -1.2 & -0.8816 & -0.4518 \\ -1.2899 & -0.6 & 1.9918 \\ -0.2914 & 0.4282 & 0 \end{bmatrix}. \tag{6.4}$$

Eigenvalues are $\eta_1 = 1.08315$, $\eta_2 = -2.0392$ and $\eta_3 = -0.84388$. For the system to be stable at an equilibrium point the eigenvalues must lie in B^{ϑ}. It can be observed from calculation that $argument(1.08315) < \frac{0.6\pi}{2}$. Thus, $\eta_1 \notin B^{\vartheta}$ implies the instability of the system at the equilibrium point Q_1.
Similarly, for equilibrium point Q_2 the Jacobian matrix is obtained as

$$J(Q_1) = \begin{bmatrix} -1.2 & -0.8816 & 0.4518 \\ -1.2899 & -0.6 & -1.9918 \\ 0.2914 & -0.4282 & 0 \end{bmatrix}, \tag{6.5}$$

with eigenvalues $\eta_1 = 1.08315$, $\eta_2 = -2.0392$ and $\eta_3 = -0.84388$. Similarly to Q_1, the equilibrium point Q_2 is also unstable since $\eta_1 \notin B^{\vartheta}$. It is worth noting that both interior equilibrium points Q_1 and Q_2 have the same eigenvalues, one of them being positive. The presence of positive real eigenvalues contradicts the conditions stated in Theorem 6.1, as the argument of the positive real numbers becomes zero. Hence, it becomes evident that the stability of the equilibrium point in the system is determined by the presence of positive real eigenvalues.

To understand the behavior of the systems, it is crucial to investigate the influence of parameters on the system dynamics. Table 6.1 presents the impact of the parameter δ on the existence of real-valued interior equilibrium points. In this analysis, the value of p is fixed at 0.3, while the parameter δ is varied within the range, -0.6 to 0.6. Previously, the results were presented for $\delta = -0.6$ in this section. Now, three different values of δ, namely -0.1, 0.1, and 0.6, are considered to evaluate the equilibrium points. It is observed that for $\delta = 0.1$, the system exhibits four interior equilibrium points. Notably, among the equilibrium points listed in Table 6.1, the interior equilibrium points for other values of δ are unstable. However, for $\delta = 0.1$, two out of the four equilibrium points are stable. This highlights the importance of studying the impact of parameters on the system behavior.

Table 6.1: Existence of Equilibrium points for different values of δ.

		Nature of stability for different ϑ			
δ	Interior equilibrium point	0.1	0.4	0.7	0.95
-0.1	$(\pm 0.8985, \mp 0.3338, -0.076)$	Unstable	Unstable	Unstable	Unstable
0.1	$(\pm 0.0868, \mp 3.4542, 0.9949)$	Unstable	Unstable	Unstable	Unstable
	$(\pm 1.1514, \mp 0.2605, 0.1161)$	Stable	Stable	Unstable	Unstable
0.6	$(\pm 0.20983, \mp 1.4296, 1.0440)$	Unstable	Unstable	Unstable	Unstable

Numerical Validation: Visualization of results is an effective way to illustrate the system dynamics over time. While mathematical results provide verification, they may not always offer a clear understanding of how the system behaves. To enhance comprehension, numerical simulations are employed to depict the asymptotic stability of the discrete fractional order Rabinovich-Fabrikant system (6.1) at the trivial equilibrium point. Specifically, the simulations consider the parameter values $\delta = -0.6$, $p = 0.3$, and $\vartheta = 0.6$, with an initial condition of $(0.4, 0.2, 0.5)$. The visualizations of these simulations can be observed in Figures 6.1 and 6.2.

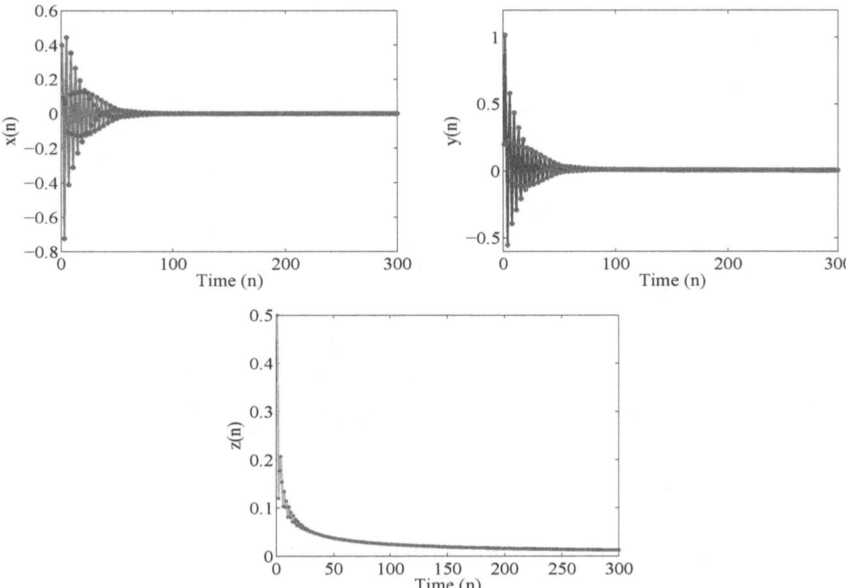

Fig. 6.1: Time series plot showing asymptotic stability of system (6.1).

From the numerical simulations of system (6.1), it can be observed that the system achieves stability after a certain period of time. During this time, oscillatory behavior is exhibited, as shown in the time series plots depicted in Figure 6.1. Moreover, the corresponding 3D phase plane diagram, presented in Figure 6.2, demonstrates that starting from the initial position, the trajectory spirals inward towards the trivial equilibrium point. As time progresses, the trajectory tends to approach the origin. It is noteworthy that the system attains stability at a fractional order of $\vartheta = 0.6$.

6.3.2 Incommensurate Fractional Order System

The conventional approach to studying fractional order systems involves assigning the same fractional order to all state variables. However, allowing different fractional orders for each state variable provides an additional degree of freedom to describe the system's essential physical properties. In order to explore these additional properties, incommensurate order systems are considered. The stability results of these systems are mathematically verified with the assistance of the following theorem.

Theorem 6.2 *[23] Let* $\alpha_\kappa = \dfrac{\omega_\kappa}{\beta_\kappa} \in (0, 1)$, *where* $\kappa = 1, 2, 3, \cdots, m$, $\omega_\kappa, \beta_\kappa > 0$ *are positive integers, the LCM of the denominators being* Φ. *The trivial solution of the system* (6.1) *with initial condition* $x(0) = x_0$ *is*

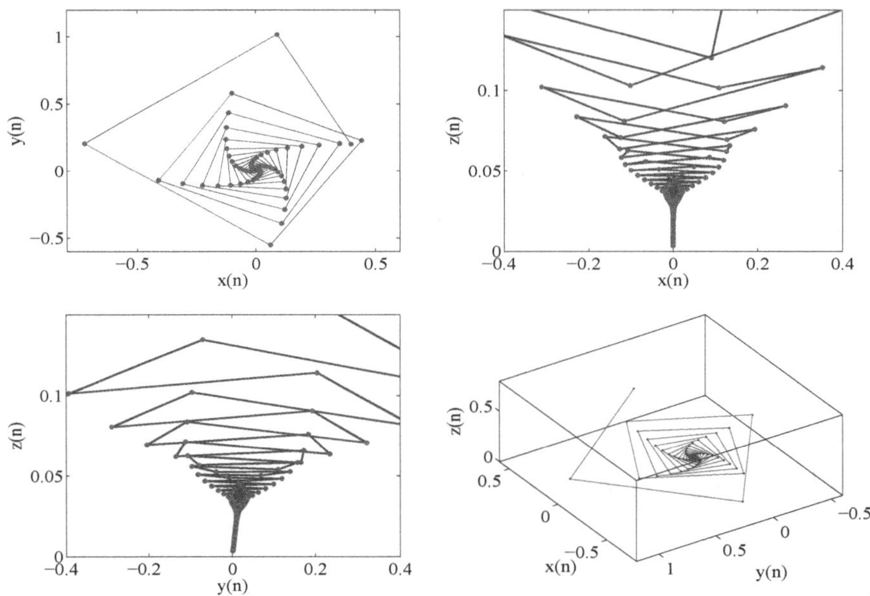

Fig. 6.2: 2D and 3D phase plane plot showing asymptotic stability of system (6.1).

asymptotically stable if the equation

$$det\left(diag\left(\eta^{\Phi\alpha_1},\eta^{\Phi\alpha_2},\ldots,\eta^{\Phi\alpha_m}\right)-\left(1-\eta^{\Phi}J\right)\right)=0, \tag{6.6}$$

has all the roots in the set $\mathbb{C}\backslash B_1^{\vartheta}$, *where* $\vartheta=\dfrac{1}{\Phi}$ *and*

$$B_1^{\vartheta}=\left\{\lambda\in\mathbb{C}:|\lambda|\leq\left(2\cos\frac{|arg\lambda|}{\vartheta}\right)^{\vartheta}\ and\ |arg\lambda|\leq\frac{\vartheta\pi}{2}\right\}.$$

Incommensurate fractional order discrete–time Rabinovich–Fabrikant system corresponding to system (6.1) takes the following form:

$$\begin{cases}\Delta^{\vartheta_1}x(n)= & y(n-1+\vartheta_1)\left(z(n-1+\vartheta_1)-1+(x(n-1+\vartheta_1))^2\right)+\delta\,x(n-1+\vartheta_1),\\ \Delta^{\vartheta_2}y(n)= & x(n-1+\vartheta_2)\left(3z(n-1+\vartheta_2)+1-(x(n-1+\vartheta_2))^2\right)+\delta y(n-1+\vartheta_2),\\ \Delta^{\vartheta_3}z(n)= & -2z(n-1+\vartheta_3)(p+x(n-1+\vartheta_3)y(n-1+\vartheta_3)),\end{cases} \tag{6.7}$$

where Δ^{ϑ_1} is the Caputo difference operator with $0<\vartheta_1,\vartheta_2,\vartheta_3\leq 1$, $n\in\mathbb{N}_{1-\vartheta_i}$, $i=1,2,3$, $\delta,p>0$ are real parameters. The fractional order discrete–time system (6.1) takes the following numerical form [22]

$$\begin{cases} x(n) = & x(0) + \dfrac{1}{\Gamma(\vartheta_1)} \sum\limits_{k=1}^{n} \dfrac{\Gamma(n-k+\vartheta_1)}{\Gamma(n-k+1)} \Big(y(j-1)\left(z(j-1) - 1 + (x(j-1))^2\right) \\ & +\delta\, x(j-1) \Big), \\ y(n) = & y(0) + \dfrac{1}{\Gamma(\vartheta_2)} \sum\limits_{k=1}^{n} \dfrac{\Gamma(n-k+\vartheta_2)}{\Gamma(n-k+1)} \Big(x(j-1)\left(3z(j-1) + 1 - (x(j-1))^2\right) \\ & +\delta y(j-1) \Big), \\ z(n) = & z(0) + \dfrac{1}{\Gamma(\vartheta_3)} \sum\limits_{k=1}^{n} \dfrac{\Gamma(n-k+\vartheta_3)}{\Gamma(n-k+1)} \left(-2z(j-1)(p + x(j-1)y(j-1))\right), \end{cases} \tag{6.8}$$

where $n = 0, 1, 2, \cdots$. Numerical verifications are conducted to compare the stability of the incommensurate order system with the commensurate order system and to explore the advantages of considering incommensurate orders. To investigate the stability of the incommensurate order discrete fractional Rabinovich-Fabrikant system (6.7), the same parameter values and initial condition as the commensurate order system (6.1) are utilized. The fractional order values chosen for numerical validation are $\vartheta_1 = 0.9$, $\vartheta_2 = 0.8$, and $\vartheta_3 = 0.4$. The stability of the incommensurate order system is established based on Theorem 6.2. The system has three equilibrium points, including a trivial equilibrium point $Q_0 = (0, 0, 0)$ and two real-valued interior equilibrium points $Q_{1,2} = (\pm 0.66395, \mp 0.45184, -0.32249)$.

Trivial Equilibrium Point Q_0: Using equation (6.6), we arrive at the following:

$$\begin{aligned} &-0.816\eta^{30} + 0.36\eta^{29} + 0.36\eta^{28} - 0.6\eta^{27} + 1.36\eta^{24} - 0.6\eta^{23} - 0.6\eta^{22} + \eta^{21} \\ &+ 2.448\eta^{20} - 0.72\eta^{19} - 0.72\eta^{18} + 0.6\eta^{17} - 2.72\eta^{14} + 0.6\eta^{13} + 0.6\eta^{12} \\ &- 2.448\eta^{10} + 0.36\eta^{9} + 0.36\eta^{8} + 1.36\eta^{4} + 0.816 = 0. \end{aligned} \tag{6.9}$$

Table 6.2: Roots of the equation (6.9).

η_1 = -1.14464	η_{16} = -0.22955+0.94959 i
η_2 = 1.14464	η_{17}= -0.33801+0.88696 i
η_3 = -0.63890-0.57491 i	η_{18} = -0.74199+0.58557 i
η_4 = 0.63890+0.57491 i	η_{19} = -0.87443+0.51951 i
η_5 = -0.47426-0.97861 i	η_{20}= -0.93633
η_6 = 0.47426+0.97861 i	η_{21} = -1.12634
η_7 = -0.47426+0.97861 i	η_{22} = -0.87443-0.51951 i
η_8 = 0.47426-0.97861i	η_{23} = -0.74199-0.58557 i
η_9 = -0.63890+0.57491 i	η_{24} = -0.33801-0.88696 i
η_{10} = 0.63890-0.57491 i	η_{25} = -0.22955-0.94959 i
η_{11} = 1.01768+0.09524 i	η_{26} = 0.29444-0.88759 i
η_{12} = 0.80508+0.51557 i	η_{27}= 0.41731 -0.96974
η_{13} = 0.90139+0.68498i	η_{28} = 0.90139-0.68498 i
η_{14} = 0.41731+0.96974 i	η_{29} = 0.80508-0.51557 i
η_{15} = 0.29444-0.94959i	η_{30}= 1.01768 -0.09524 i

Table 6.3: Roots of the equation (6.10).

η_1 = 0.92275	η_{16} = -0.94855+0.04194 i
η_2 = 1.06098	η_{17}= -1.09878
η_3 = 1.14090	η_{18} = -0.94855-0.04194 i
η_4 = 0.86395+0.56505 i	η_{19} = -0.78944-0.52539 i
η_5 = 0.77048+0.50732 i	η_{20}= -0.73675 -0.57358 i
η_6 = 0.81151+0.70976 i	η_{21} = -0.85248-0.67230 i
η_7 = 0.41512+0.90659 i	η_{22} = -0.36891 -0.88607 i
η_8 = 0.34709+0.91470 i	η_{23} = -0.26858-0.91845 i
η_9 = 0.22077+0.98060 i	η_{24} = -0.24833-1.04029 i
η_{10} = -0.24833+1.04029 i	η_{25} = 0.22077-0.98060 i
η_{11} = -0.26858+0.91845 i	η_{26} = 0.34709-0.91470 i
η_{12} = -0.36891 +0.88607 i	η_{27}= 0.41512 -0.90659 i
η_{13} = -0.85248+0.67230 i	η_{28} = 0.81151-0.70976 i
η_{14} = -0.73675 +0.57358 i	η_{29} = 0.77048-0.50732 i
η_{15} = -0.78944+0.52539 i	η_{30}= 0.86395-0.56505 i

It is clear that
$|\eta_l| > (2\cos(10\,|arg\eta_l|))^{0.1}$ for $l \in \{1,2,5,6,7,8,11,13,14,21,27,28,30\}$ and
$|arg\eta_l| > \frac{\pi}{20}$ for $l \in \{1,3,4,5,6,7,8,9,10,12,13,14,15,16,17,18,19,20,21,22,$
$23,24,25,26,27,28,29\}$. From Theorem 6.2 that $\eta_l \in \mathbb{C}\backslash B_1^{\vartheta}$ for $1 \leq l \leq 30$, ensuring the asymptotic stability of the system at the trivial equilibrium point.

Real Valued Interior Equilibrium Points $Q_{1,2}$: The characteristic equation at equilibrium point $Q_{1,2}$ is given by

$$\begin{aligned}&1.8640\eta^{30} - 0.8529\eta^{29} - 0.13168\eta^{28} - 0.41733\eta^{24} - 0.6\eta^{23} - 1.2\eta^{22} + \eta^{21}\\&- 5.5920\eta^{20} + 1.7059\eta^{19} + 0.26336\eta^{18} + 0.8346\eta^{14} + 0.6\eta^{13} + 1.2\eta^{12}\\&+ 5.5920\eta^{10} - 0.8529\eta^{9} - 0.1316\eta^{8} - 0.4173\eta^{4} - 1.8640 = 0.\end{aligned} \tag{6.10}$$

Roots of the equation (6.10) are given in Table 6.3 From Theorem (6.2), it can be confirmed that the eigenvalues $\eta_{1,2} \in B_1^{\vartheta}$. i.e. $|\eta_{1,2}| < (2\cos(10\,|arg\eta_l|))^{0.1}$ and $|arg\eta_{1,2}| < \frac{\pi}{20}$. Thus, it is evident that the system (6.7) is unstable at the equilibrium points $Q_{1,2}$.

The numerical simulations presented in Figures 6.3 and 6.4 clearly demonstrate the convergence of system orbits towards the origin. When comparing with the commensurate order system, it can be observed that for the same parameter values, the incommensurate order system achieves stability with fractional orders $\vartheta_1 = 0.9$, $\vartheta_2 = 0.8$, and $\vartheta_3 = 0.4$. This indicates that the different physical behavior of the system, resulting from the use of incommensurate orders, contributes to its stability. The transition of the systems from stability to periodic oscillations and eventually to aperiodic oscillations will be further explored in the next section through bifurcation analysis, which holds great significance.

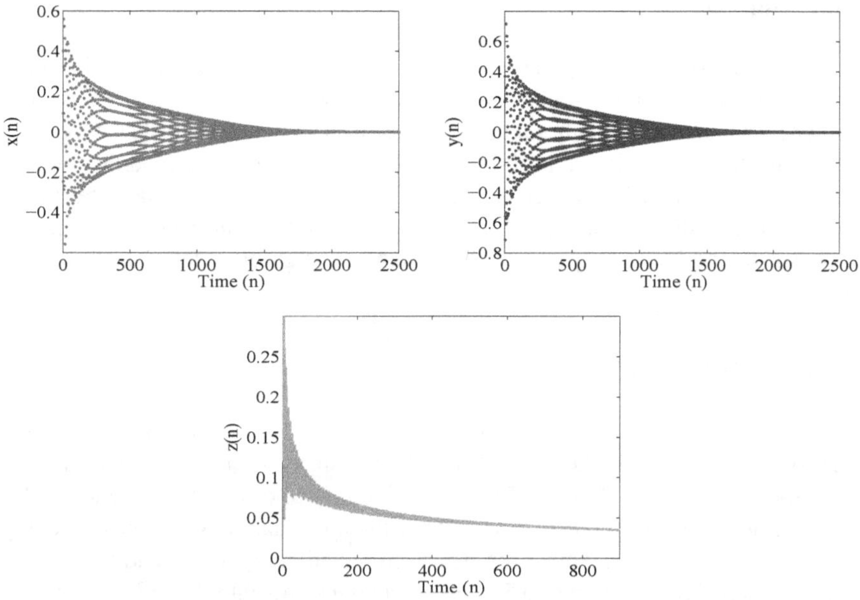

Fig. 6.3: Time series plot showing asymptotic stability of system (6.7).

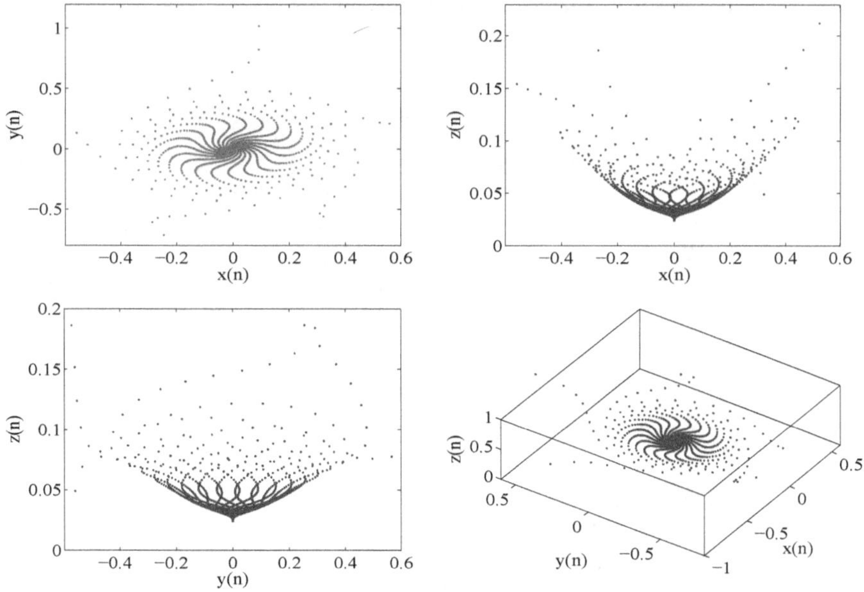

Fig. 6.4: 2D and 3D phase plane plot showing asymptotic stability of system (6.7).

6.4 Bifurcation Analysis

Bifurcation analysis is a powerful tool that allows us to understand how the qualitative behavior of a system changes abruptly due to variations in its parameters. It not only provides insights into the system's transition between different states, but also sheds light on the influence of parameters on its behavior and dynamics. By studying bifurcations, we can develop control strategies and make appropriate modifications based on the specific requirements of the system. Consequently, the studies of bifurcation and chaos theories holds great significance in numerous interdisciplinary fields. In this context, we will investigate the chaotic behavior of both the commensurate and incommensurate order systems through bifurcation analysis and Lyapunov exponents.

6.4.1 Commensurate Order System

To explore the chaotic behavior of the discrete fractional order Rabinovich-Fabrikant system (6.1), we vary the fractional order parameter ϑ within the range (0,1). The parameter values $\delta = -0.65$ and $p = 0.01$ are fixed, along with initial conditions of (0.05, -0.05, 0.1). The presence of chaotic behavior in the system is confirmed through the calculation of Lyapunov exponents. Bifurcation diagrams and corresponding attractors are obtained for different values of ϑ, and these results are presented in Figure 6.5 and Figure 6.6.

The bifurcation diagrams shown in Figure 6.5 offer valuable insights into the influence of the fractional order parameter on the dynamics of the system. For small fractional orders, the system exhibits chaotic behavior. As the fractional order increases, the system transition's into an unstable region characterized by the formation of limit cycles. Within the fractional order range of (0.5, 0.85), the system stabilizes and enters a phase of periodic orbits for the state variables x and y. Beyond this range, the system undergoes another

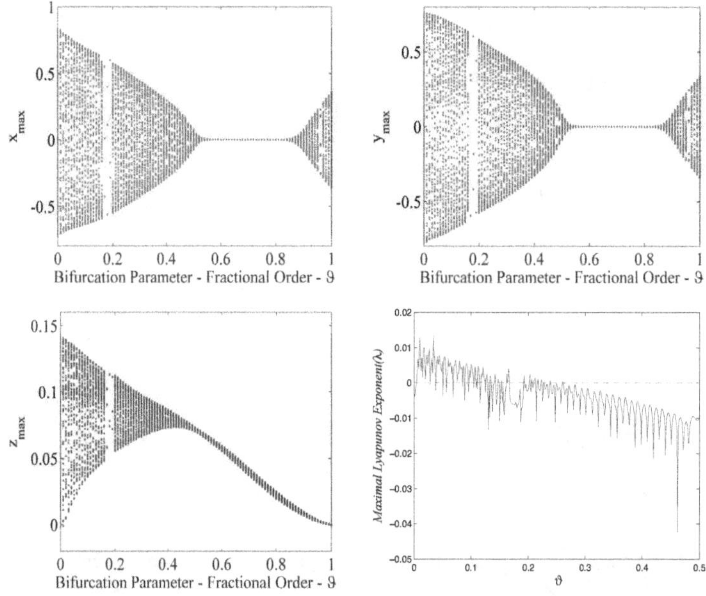

Fig. 6.5: Bifurcation diagram and Lyapunov exponent of system (6.1) varying $\vartheta \in (0, 1)$.

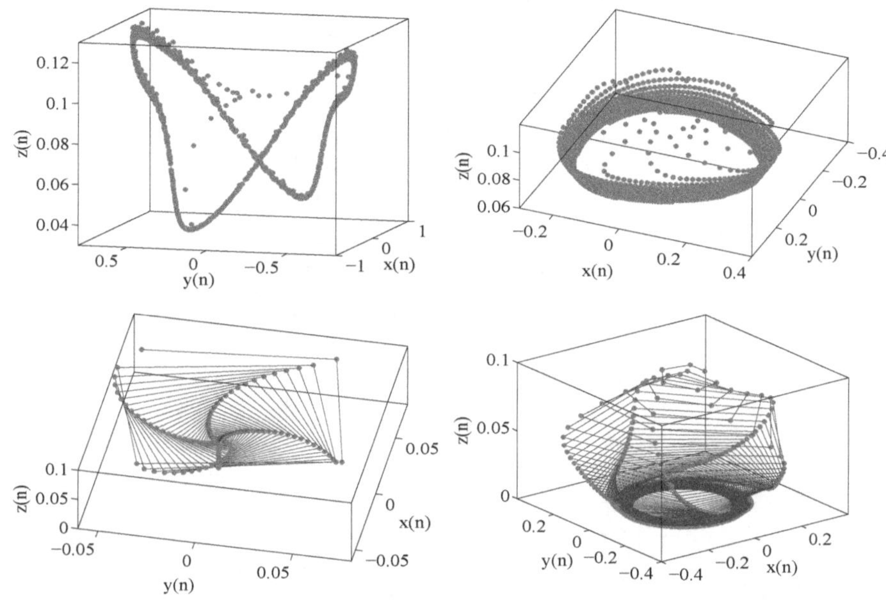

Fig. 6.6: 3D attractors corresponding to Figure 6.5 for $\vartheta = 0.1, 0.4, 0.7, 0.95$.

transition, returning to a state of periodic orbits. These distinct phases of the system are further visualized through 3-D attractors in Figure 6.6.

6.4.2 Approximate Entropy

The approximate entropy proposed in [20] is a useful tool in representing the randomness of the time series in a dynamical system. The approximate entropy calculations are performed as follows.

$$ApEn(h, r, W) = \Theta_{h+1}(r) - \Theta_h(r), \tag{6.11}$$

where h is the number of data points to be compared, the tolerance factor is represented by r and the total time series data is W. Here, $\Theta_h(r) = \frac{1}{W-h+1} \sum_{j=1}^{W-h+1} \log M_j^h(r)$, where the value of $M_j^h(r) = \frac{|\chi(k) - \chi(l)|}{W-h+1}$, $\chi(k)$ is the kth block of time series sequences. The complexity analysis is performed for $\delta = -0.65$ and $p = 0.2$ for varying fractional orders $\vartheta = 0.1, 0.3, 0.5$ with different tolerance values. The values are tabulated in Table 6.4 and plotted against tolerance (r) in Figure 6.7.

6.4.3 Incommensurate Order Systems

To explore the chaotic behavior of the Rabinovich–Fabrikant system under incommensurate fractional orders, numerical simulations are conducted with specific parameter values. The parameters are set as $\delta = -1$, $p = -0.1$, $\vartheta_1 = 0.5$, and $\vartheta_3 = 0.1$. The variable ϑ_2 is varied between 0 and 1 in order to investigate its impact on the system dynamics.

Table 6.4: Approximate entropy of identical order Rabinovich-Fabrikant system.

p	Tolerance (r)	Fractional Order (ϑ)		
		$\vartheta = 0.1$	$\vartheta = 0.3$	$\vartheta = 0.5$
0.01	$r = 0.2$	0.3197	0.2551	0.1920
	$r = 0.3$	0.2657	0.1931	0.1582
	$r = 0.5$	0.1992	0.1416	0.1719
	$r = 0.7$	0.2146	0.1366	0.1992
	$r = 0.9$	0.2635	0.1851	0.2301
0.2	$r = 0.2$	0.2527	0.1616	0.0921
	$r = 0.3$	0.2233	0.1288	0.1095
	$r = 0.5$	0.1728	0.1158	0.1136
	$r = 0.7$	0.1785	0.0979	0.1076
	$r = 0.9$	0.2346	0.1136	0.0874

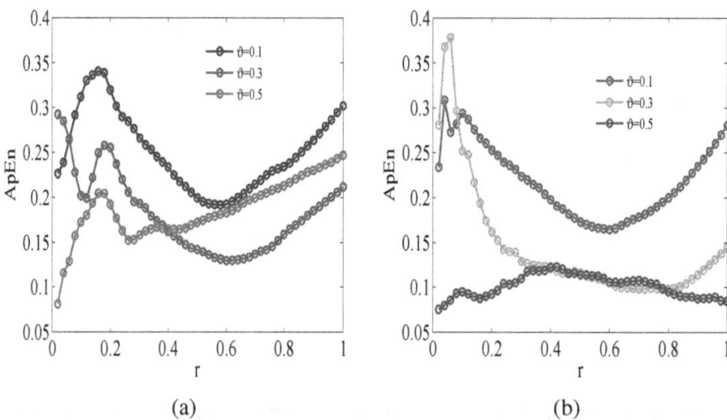

Fig. 6.7: 2-D plot of approximate entropy with varying tolerance (a) $p = 0.01$, $W = 1500$, (b) $p = 0.2$, $W = 1500$.

The initial conditions are chosen as $(0.05, -0.05, 0.01)$. The simulations are complemented by the analysis of Lyapunov exponents, and the resulting attractors at different values of ϑ_2 are presented in Figure 6.8 and Figure 6.9. These visualizations provide valuable insights into the chaotic behavior exhibited by the system under incommensurate fractional orders. The investigation focuses on the transition between different states in the incommensurate order system, specifically in relation to the fractional order ϑ_2 of the second state variable. By keeping the orders of the other two state variables fixed and varying only the fractional order of the second state variable, the study examines how the system responds to this change. It is observed that the system initially exhibits chaotic behavior for smaller values of ϑ_2, but transitions to a state of stability for larger values. This highlights the significance of properly considering and treating the individual state variables in the system analysis.

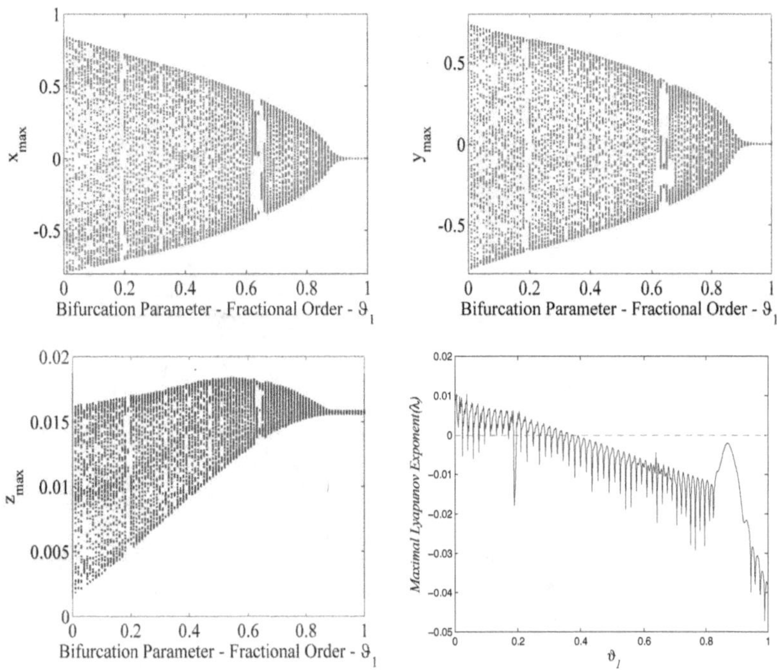

Fig. 6.8: Bifurcation diagram and Lyapunov exponent of system (6.7) varying $\vartheta \in (0, 1)$.

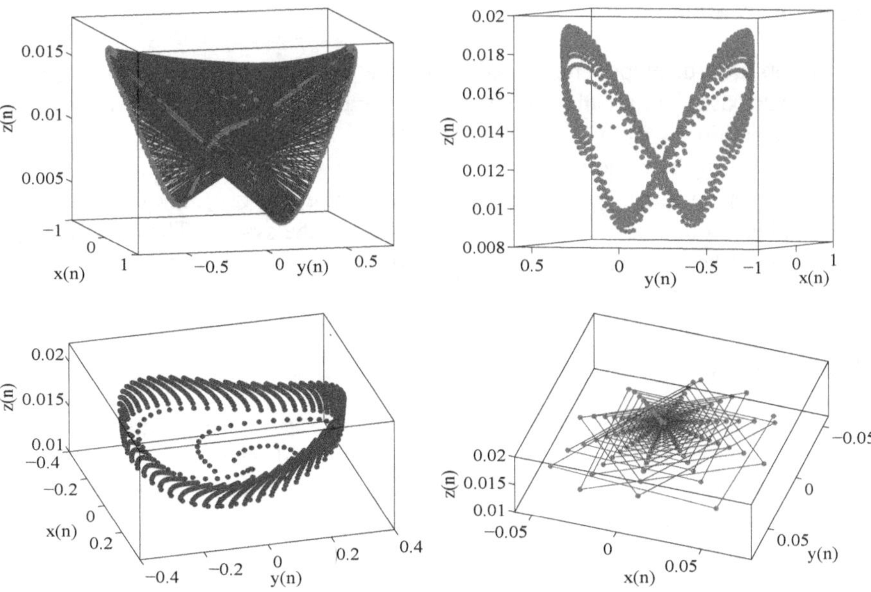

Fig. 6.9: 3D attractors corresponding to Figure 6.8 for $\vartheta_2 = 0.1, 0.4, 0.7, 0.95$.

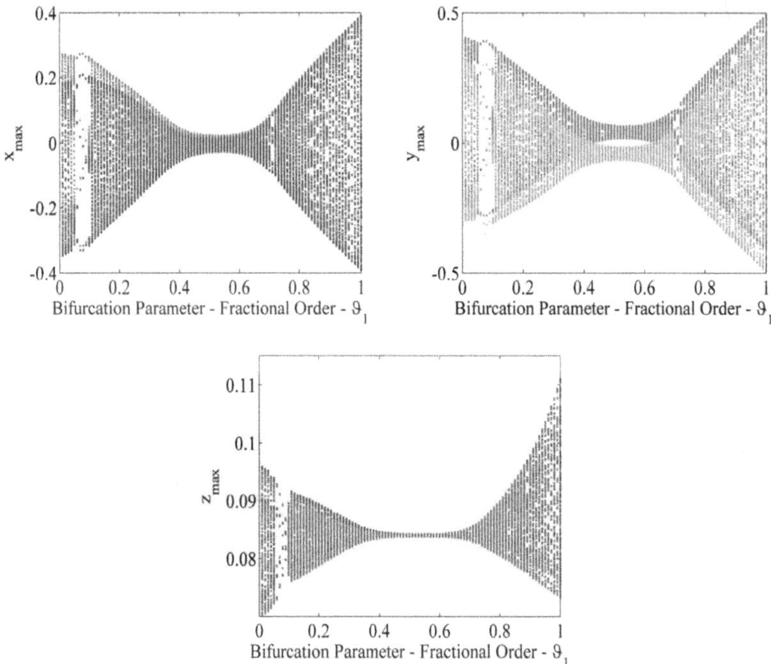

Fig. 6.10: Coexisting bifurcation diagram of system (6.7) varying $\vartheta \in (0, 1)$.

6.4.4 Coexisting bifurcations

The coexisting attractors are crucial for comprehending how the dynamics of the system change with different initial conditions. The coexisting bifurcation is obtained for parameter values $\delta = -0.1$, $p = 0.07$ $\vartheta_1 = 0.15$, $\vartheta_3 = 0.05$ with initial conditions $(0.1, 0.1, 0.1)$ and $(-0.1, 0.1, 0.1)$ varying ϑ_2 between 0 and 1. Figure 6.10 illustrates the impact of changes in the initial conditions on the system dynamics. Among the three state variables, $y(n)$ undergoes a more significant change compared to $x(n)$ and $z(n)$. The corresponding attractors provide valuable insights into the behavior of the chaotic state of the system. The coexistence of different attractors is crucial for understanding the sensitivity of the system to initial conditions and for making appropriate adjustments to improve its performance.

6.4.5 Complexity of Chaotic Series

This section presents the complexity measure of the chaotic time series through approximate entropy calculations obtained from (6.11). The complexity analysis is performed for $\delta = -1$ and $p = -0.1$ for fixed $\vartheta_1 = 0.5$, $\vartheta_3 = 0.3$ and varying fractional orders, $\vartheta_2 = 0.1, 0.3, 0.5$ with different tolerance values. The values are tabulated in Table 6.5 and plotted against tolerance (r) in Figure 6.12.

6.5 Complex Rabinovich–Fabrikant System

The exploration of system state variables in the complex domain has gained recent interest and has shown a wide range of applications. Introducing real and imaginary parts to the system provides a novel approach to

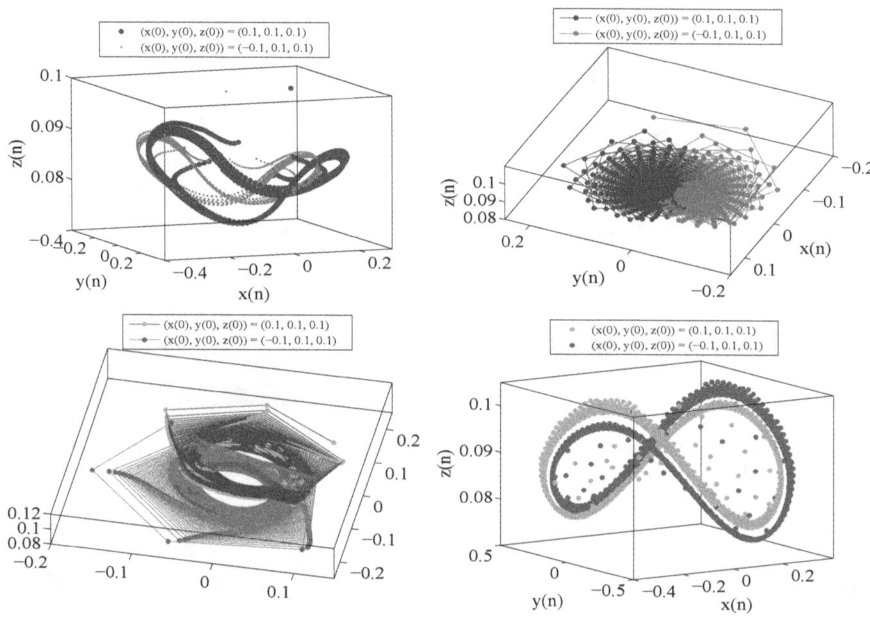

Fig. 6.11: 3D attractors corresponding to Figure 6.10 for $\vartheta_2 = 0.1, 0.4, 0.7, 0.95$.

Table 6.5: Approximate entropy of non-identical order Rabinovich-Fabrikant system.

		Fractional Order (ϑ)		
ϑ_3	Tolerance (r)	$\vartheta_2 = 0.1$	$\vartheta_2 = 0.3$	$\vartheta_2 = 0.5$
	$r = 0.2$	0.2500	0.2663	0.2433
	$r = 0.3$	0.2722	0.2790	0.1961
0.1	$r = 0.5$	0.2437	0.1884	0.1343
	$r = 0.7$	0.2669	0.1794	0.1191
	$r = 0.9$	0.3508	0.2680	0.1757
	$r = 0.2$	0.2567	0.2682	0.4115
	$r = 0.3$	0.2727	0.2643	0.3166
0.3	$r = 0.5$	0.2414	0.1763	0.1944
	$r = 0.7$	0.2621	0.1787	0.1872
	$r = 0.9$	0.3496	0.2673	0.2332

studying systems in different dimensions. Converting the system to the complex domain effectively doubles the number of state variables. While analyzing the complex form of the system may require more time, the output of the complex system is valuable and enhances our understanding to a greater extent. Although there have been various complex systems proposed, ranging from Lorentz and Rossler's systems to Chua systems, there is a lack of complex systems involving discrete fractional order in the existing literature. Therefore, this section introduces the discrete fractional Rabinovich-Fabrikant system with complex state variables and investigates its chaotic behavior through the use of bifurcation diagrams and attractors. Setting $x = w_1 + i\, w_2$, $y = w_3 + i\, w_4$ and $z = w_5 + i\, w_6$, where $i = \sqrt{-1}$ in system (6.1) and separating the real and imaginary parts,

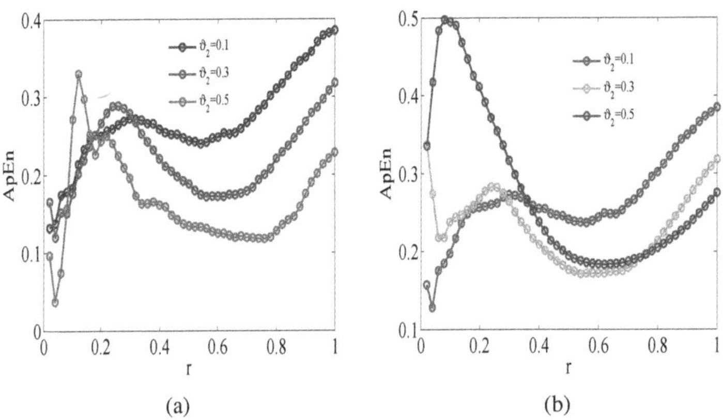

Fig. 6.12: 2-D plot of approximate entropy with varying tolerance (a) $\vartheta_3 = 0.1$, $W = 1500$, (b) $\vartheta_3 = 0.3$, $W = 1500$.

we obtain,

$$
\left\{
\begin{array}{ll}
\Delta^{\vartheta} w_1(n) = & w_3(n-1+\vartheta)\left(w_5(n-1+\vartheta) - 1 + (w_1(n-1+\vartheta))^2 - (w_2(n-1+\vartheta))^2\right) \\
& -w_4(n-1+\vartheta)\left(w_6(n-1+\vartheta) + 2w_1(n-1+\vartheta)w_2(n-1+\vartheta)\right) \\
& +\delta\, w_1(n-1+\vartheta), \\
\Delta^{\vartheta} w_2(n) = & w_4(n-1+\vartheta)\left(w_5(n-1+\vartheta) - 1 + (w_1(n-1+\vartheta))^2 - (w_2(n-1+\vartheta))^2\right) \\
& +w_3(n-1+\vartheta)\left(w_6(n-1+\vartheta) + 2w_1(n-1+\vartheta)w_2(n-1+\vartheta)\right) \\
& +\delta\, w_2(n-1+\vartheta), \\
\Delta^{\vartheta} w_3(n) = & w_1(n-1+\vartheta)\left(3w_5(n-1+\vartheta) + 1 - (w_1(n-1+\vartheta))^2 + (w_2(n-1+\vartheta))^2\right) \\
& -w_2(n-1+\vartheta)\left(3w_6(n-1+\vartheta) - 2w_1(n-1+\vartheta)w_2(n-1+\vartheta)\right) \\
& +\delta\, w_3(n-1+\vartheta), \\
\Delta^{\vartheta} w_4(n) = & w_2(n-1+\vartheta)\left(3w_5(n-1+\vartheta) + 1 - (w_1(n-1+\vartheta))^2 + (w_2(n-1+\vartheta))^2\right) \\
& +w_1(n-1+\vartheta)\left(3w_6(n-1+\vartheta) - 2w_1(n-1+\vartheta)w_2(n-1+\vartheta)\right) \\
& +\delta\, w_4(n-1+\vartheta), \\
\Delta^{\vartheta} w_5(n) = & -2w_5(n-1+\vartheta)\Bigg(p + w_1(n-1+\vartheta)w_3(n-1+\vartheta) - w_2(n-1+\vartheta) \\
& w_4(n-1+\vartheta)\Bigg) + 2w_6(n-1+\vartheta)\Bigg(w_1(n-1+\vartheta)w_4(n-1+\vartheta) \\
& +w_2(n-1+\vartheta)w_3(n-1+\vartheta)\Bigg), \\
\Delta^{\vartheta} w_6(n) = & -2w_6(n-1+\vartheta)\Bigg(p + w_1(n-1+\vartheta)w_3(n-1+\vartheta) - w_2(n-1+\vartheta) \\
& w_4(n-1+\vartheta)\Bigg) - 2w_5(n-1+\vartheta)\Bigg(w_1(n-1+\vartheta)w_4(n-1+\vartheta) \\
& +w_2(n-1+\vartheta)w_3(n-1+\vartheta)\Bigg),
\end{array}
\right. \tag{6.12}
$$

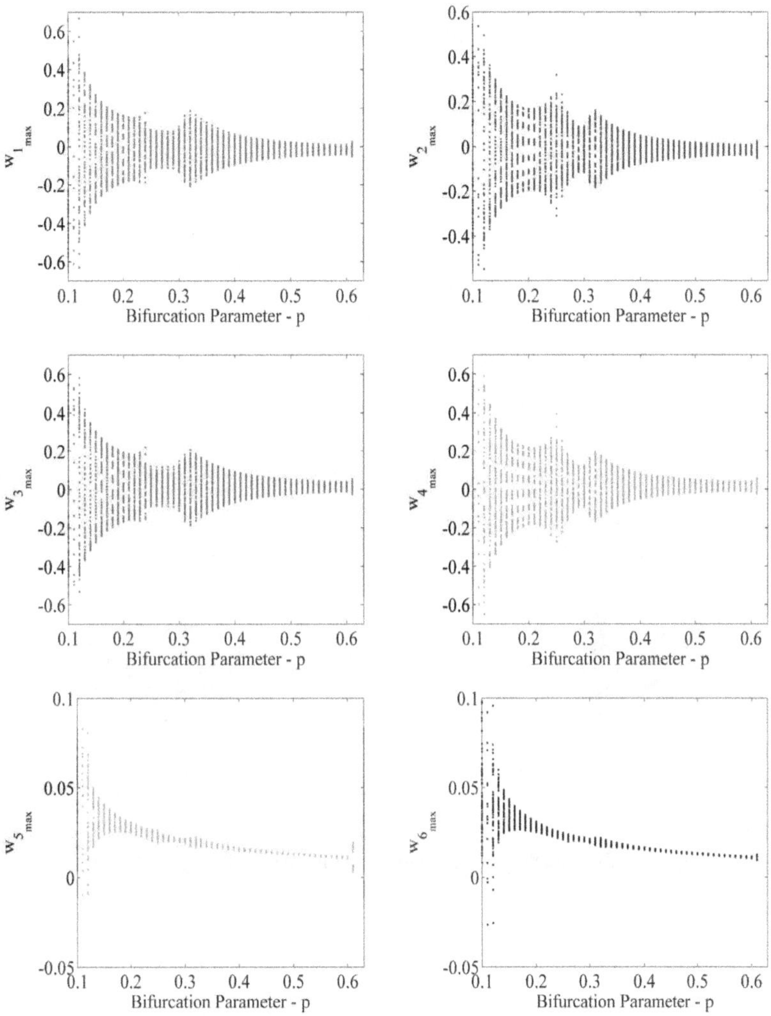

Fig. 6.13: Bifurcation diagram of system (6.12) varying $p \in (0, 0.62)$.

where $n \in \mathbb{N}_{1-\vartheta}$. Bifurcation analysis is conducted for the commensurate order complex Rabinovich-Fabrikant system (6.12) by considering $\delta = 0.1$ and $\vartheta = 0.3$. The parameter p is varied in the range $(0, 0.62)$, and the results are presented in Figure 6.13, with Lyapunov exponents shown in Figure 6.14. The corresponding attractors are illustrated in Figure 6.15 and Figure 6.16. In the case of a system defined with complex variables, chaotic behavior is observed for values of p close to zero. As the values of p tend to 1, the system undergoes a transition from periodic orbits to stable orbits. The attractors are obtained by considering the real and imaginary parts of the complex terms separately in a three-dimensional space.

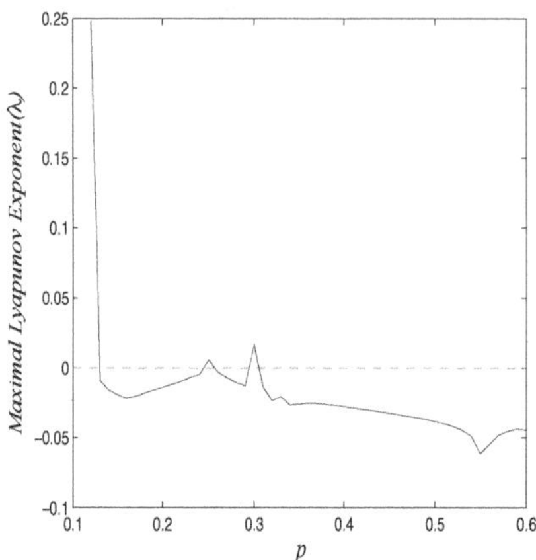

Fig. 6.14: Lyapunov exponent of system (6.12) varying $p \in (0, 0.62)$ corresponding to Figure 6.13.

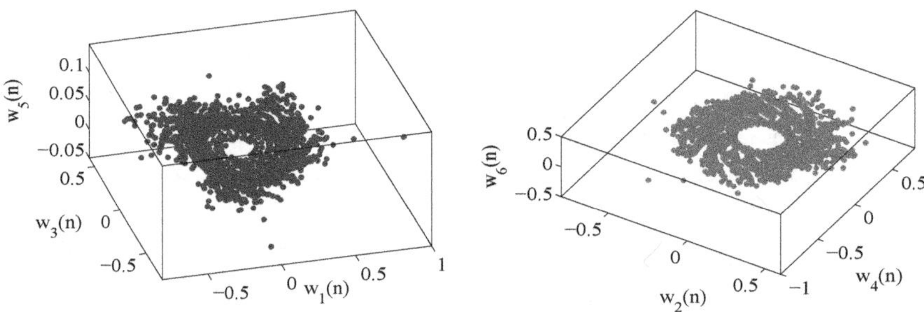

Fig. 6.15: 3D attractors corresponding to Figure 6.10 for $p = 0.1$.

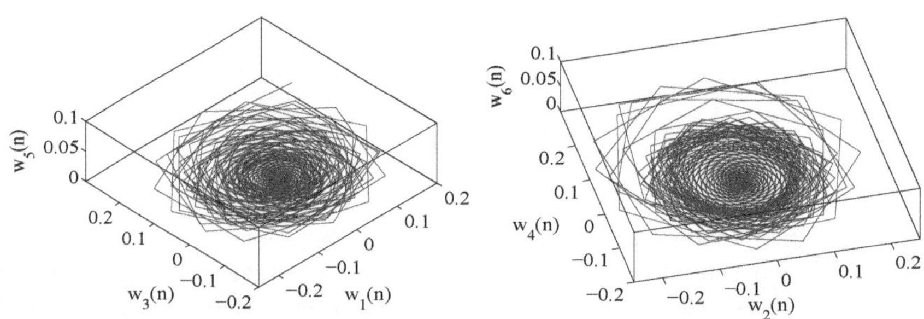

Fig. 6.16: 3D attractors corresponding to Figure 6.10 for $p = 0.35$.

6.5.1 Randomness of Chaotic Series

This section presents the complexity measure of the chaotic time series through approximate entropy calculations obtained from (6.11). The analysis of complexity is performed for $\delta = 0.1$ and $p = 0.2$ for varying

Table 6.6: Approximate entropy of identical order complex Rabinovich-Fabrikant system.

p	Tolerance (r)	Fractional Order (ϑ)		
		$\vartheta = 0.1$	$\vartheta = 0.3$	$\vartheta = 0.5$
	$r = 0.2$	0.4319	0.3458	0.2843
	$r = 0.3$	0.3005	0.2528	0.3934
0.15	$r = 0.5$	0.1835	0.1676	0.4397
	$r = 0.7$	0.1366	0.1286	0.3955
	$r = 0.9$	0.1367	0.1495	0.4371
	$r = 0.2$	0.3959	0.3364	0.1502
	$r = 0.3$	0.3015	0.2566	0.2574
0.2	$r = 0.5$	0.1857	0.1708	0.3293
	$r = 0.7$	0.1411	0.1367	0.3396
	$r = 0.9$	0.1566	0.1936	0.4723

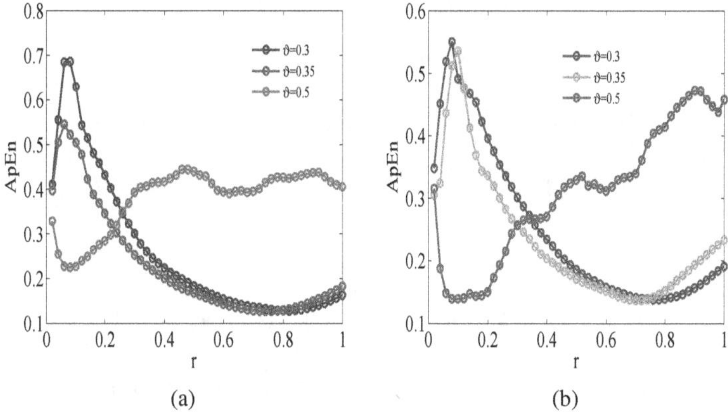

Fig. 6.17: 2-D plot of approximate entropy with varying tolerance (a) $p = 0.15$, $W = 3000$, (b) $p = 0.2$, $W = 3000$.

fractional orders $\vartheta = 0.3, 0.35, 0.5$ with different tolerance values. The values are tabulated in Table 6.6 and plotted against tolerance (r) in Figure 6.17.

6.6 Incommensurate Order Complex Rabinovich–Fabrikant System

The incommensurate order complex Rabinovich – Fabrikant system takes the following form:

$$
\left\{
\begin{aligned}
\Delta^{\vartheta_1} w_1(n) = \ & w_3(n-1+\vartheta_1)\Big(w_5(n-1+\vartheta_1) - 1 + (w_1(n-1+\vartheta_1))^2 \\
& -(w_2(n-1+\vartheta_1))^2\Big) - w_4(n-1+\vartheta_1)\Big(w_6(n-1+\vartheta_1) \\
& +2w_1(n-1+\vartheta_1)w_2(n-1+\vartheta_1)\Big) + \delta\, w_1(n-1+\vartheta_1), \\
\Delta^{\vartheta_1} w_2(n) = \ & w_4(n-1+\vartheta_1)\Big(w_5(n-1+\vartheta_1) - 1 + (w_1(n-1+\vartheta_1))^2 \\
& -(w_2(n-1+\vartheta_1))^2\Big) + w_3(n-1+\vartheta_1)\Big(w_6(n-1+\vartheta_1) \\
& +2w_1(n-1+\vartheta_1)w_2(n-1+\vartheta_1)\Big) + \delta\, w_2(n-1+\vartheta_1), \\
\Delta^{\vartheta_2} w_3(n) = \ & w_1(n-1+\vartheta_2)\Big(3w_5(n-1+\vartheta_2) + 1 - (w_1(n-1+\vartheta_2))^2 \\
& +(w_2(n-1+\vartheta_2))^2\Big) - w_2(n-1+\vartheta_2)\Big(3w_6(n-1+\vartheta_2) \\
& -2w_1(n-1+\vartheta_2)w_2(n-1+\vartheta_2)\Big) + \delta\, w_3(n-1+\vartheta_2), \\
\Delta^{\vartheta_2} w_4(n) = \ & w_2(n-1+\vartheta_2)\Big(3w_5(n-1+\vartheta_2) + 1 - (w_1(n-1+\vartheta_2))^2 \\
& +(w_2(n-1+\vartheta_2))^2\Big) + w_1(n-1+\vartheta_2)\Big(3w_6(n-1+\vartheta_2) \\
& -2w_1(n-1+\vartheta_2)w_2(n-1+\vartheta_2)\Big) + \delta\, w_4(n-1+\vartheta_2), \\
\Delta^{\vartheta_3} w_5(n) = \ & -2w_5(n-1+\vartheta_3)\Big(p + w_1(n-1+\vartheta_3)w_3(n-1+\vartheta_3) \\
& -w_2(n-1+\vartheta_3)w_4(n-1+\vartheta_3)\Big) + 2w_6(n-1+\vartheta_3) \\
& (w_1(n-1+\vartheta_3)w_4(n-1+\vartheta_3) + w_2(n-1+\vartheta_3)w_3(n-1+\vartheta_3)), \\
\Delta^{\vartheta_3} w_6(n) = \ & -2w_6(n-1+\vartheta_3)\Big(p + w_1(n-1+\vartheta_3)w_3(n-1+\vartheta_3) \\
& -w_2(n-1+\vartheta_3)w_4(n-1+\vartheta_3)\Big) - 2w_5(n-1+\vartheta_3) \\
& (w_1(n-1+\vartheta_3)w_4(n-1+\vartheta_3) + w_2(n-1+\vartheta_3)w_3(n-1+\vartheta_3)),
\end{aligned}
\right. \tag{6.13}
$$

where $n \in \mathbb{N}_{1-\vartheta_i}, i = 1, 2, 3$. The incommensurate order complex Rabinovich-Fabrikant system (6.13) is analyzed with the parameter values $\delta = -0.1$, $p = 0.24$, $\vartheta_2 = 0.5$, $\vartheta_3 = 0.2$, and by varying the parameter $\vartheta_1 \in (0, 0.6)$. The bifurcation analysis is performed, and the simulations are presented in Figure 6.18, while the corresponding Lyapunov exponents are shown in Figure 6.19. In the case of the incommensurate order system, all six state variables exhibit chaotic behavior for fractional order ϑ_1 between 4.5 and 5.2.

The attractors corresponding to the bifurcation diagram presented in Figure 6.18 are visualized in Figures 6.20 and 6.21. In Figure 6.20, the bifurcation diagram for initial values of ϑ_1 appears as a straight line, indicating the stable behavior of the system. This is further confirmed by the spiral trajectory observed in the attractor plot, where the system moves inward towards the equilibrium point. However, for higher values of

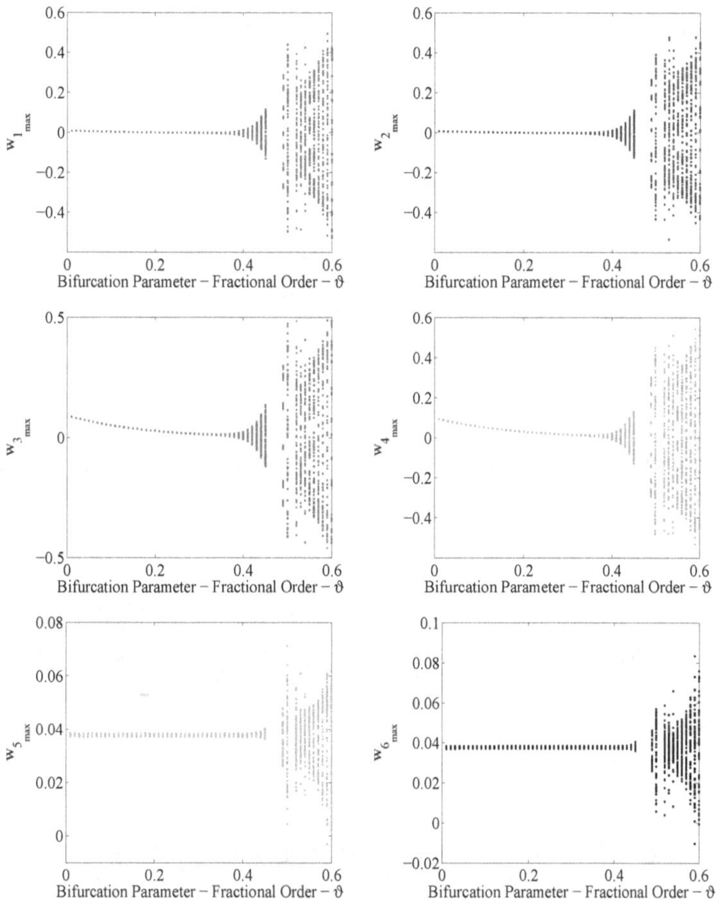

Fig. 6.18: Bifurcation diagram of system (6.13) varying $\vartheta \in (0, 0.6)$.

the fractional order ϑ_1, chaotic behavior is observed, as depicted in Figure 6.21. The formation of a limit cycle can be observed, indicating the transition of the system from a periodic state to an aperiodic state and then back to a periodic state. These findings are consistent with the bifurcation diagrams and the corresponding Lyapunov exponent plot shown in Figure 6.19.

6.6.1 Randomness of Chaotic Series

This section presents the complexity measure of the chaotic time series through approximate entropy calculations obtained from (6.11). The analysis of complexity is performed for $\delta = -0.1$ and $p = 0.3$ for fixed $\vartheta_2 = 0.5$, $\vartheta_3 = 0.2$ and varying fractional orders $\vartheta_1 = 0.15, 0.25, 0.35$ with different tolerance values. The values are tabulated in Table 6.7 and plotted against tolerance (r) in Figure 6.22.

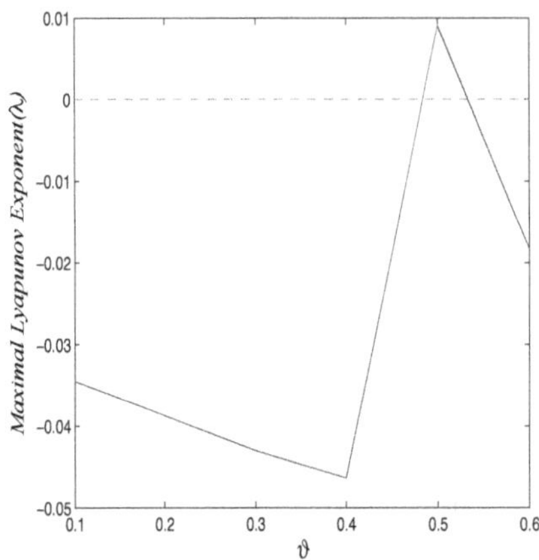

Fig. 6.19: Lyapunov exponent of system (6.13) varying $\vartheta \in (0, 0.6)$ corresponding to Figure 6.18.

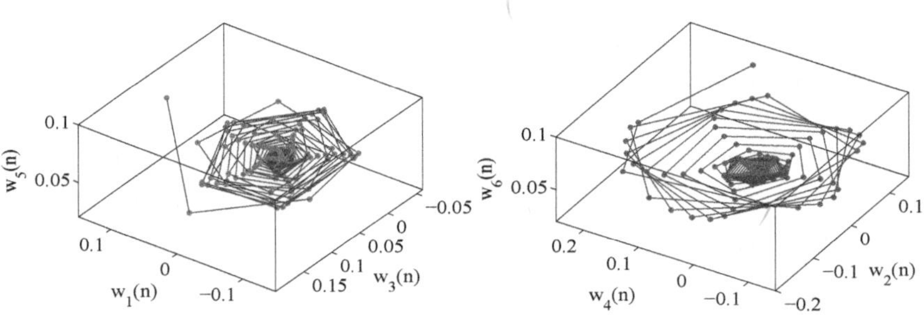

Fig. 6.20: 3D attractors corresponding to Figure 6.18 for $\vartheta = 0.3$.

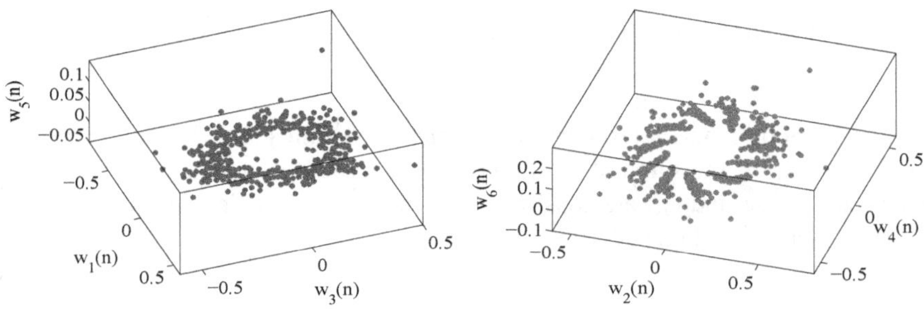

Fig. 6.21: 3D attractors corresponding to Figure 6.18 for $\vartheta = 0.55$.

6.7 Summary

The chapter introduces the discrete-time fractional order Rabinovich-Fabrikant system with cubic and quadratic nonlinearities, considering both real and complex state variables. The mathematical analysis is divided into two parts: one focusing on the commensurate order system and the other on the incommensurate

Table 6.7: Approximate entropy of non-identical order complex Rabinovich-Fabrikant system.

		Fractional Order (ϑ)		
p	Tolerance (r)	$\vartheta_1 = 0.1$	$\vartheta_1 = 0.3$	$\vartheta_1 = 0.5$
0.24	$r = 0.2$	0.1446	0.1749	0.2585
	$r = 0.3$	0.1338	0.1863	0.2703
	$r = 0.5$	0.1313	0.1992	0.2310
	$r = 0.7$	0.1273	0.1859	0.2126
	$r = 0.9$	0.1166	0.1707	0.1767
0.3	$r = 0.2$	0.1027	0.1544	0.1820
	$r = 0.3$	0.1152	0.1617	0.2133
	$r = 0.5$	0.1154	0.1523	0.2163
	$r = 0.7$	0.1139	0.1365	0.1873
	$r = 0.9$	0.1075	0.1286	0.1524

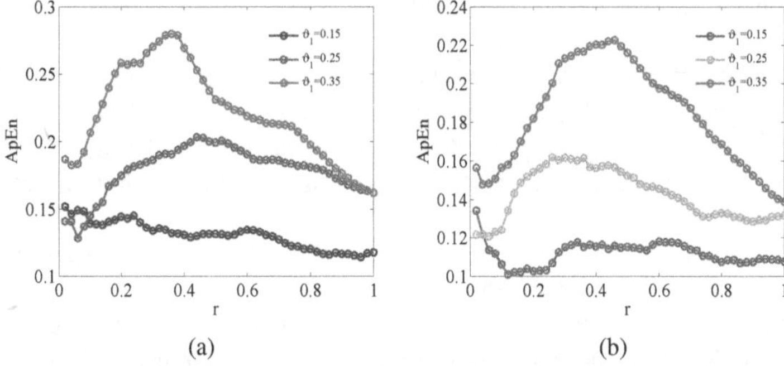

Fig. 6.22: 2-D plot of approximate entropy with varying tolerance (a) $p = 0.24$, $W = 3000$, (b) $p = 0.3$, $W = 3000$.

order system. The qualitative behaviors of the systems, such as stability, formation of limit cycles, and state transitions, are explored by investigating the influence of the fractional orders. A comparison between the commensurate and incommensurate order systems is presented, highlighting the advantages of the incommensurate order system with its distinct fractional orders assigned to each state variable. The inclusion of distinct fractional orders for each equation of the system helps reveal previously unidentified physical features. In the second part of the chapter, the Rabinovich-Fabrikant system with complex state variables is introduced for the first time in the context of fractional order discrete-time calculus. The chaotic behavior of the complex system is illustrated through bifurcation diagrams, Lyapunov exponents, and attractors. This chapter serves as a starting point for the construction and study of complex chaotic systems, providing valuable insights into their qualitative properties. Future extensions of the chapter could explore variable order complex systems, where the fractional order varies as a function of time, opening up new avenues for research in this field.

References

[1] Cun-Zheng Ning and Hermann Haken. Detuned lasers and the complex Lorenz equations: Subcritical and supercritical Hopf bifurcations. Physical Review A 41, no. 7 (1990): 3826.

[2] Gamal M. Mahmoud and Tassos Bountis. The dynamics of systems of complex nonlinear oscillators: A review. International Journal of Bifurcation and Chaos 14, no. 11 (2004): 3821–3846.

[3] Shutang Liu and Fangfang Zhang. Complex function projective synchronization of complex chaotic system and its applications in secure communication. Nonlinear Dynamics 76 (2014): 1087–1097.

[4] Xiangjun Wu, Changjiang Zhu and Haibin Kan. An improved secure communication scheme based passive synchronization of hyperchaotic complex nonlinear system. Applied Mathematics and Computation 252 (2015): 201–214.

[5] Leyuan Wang, Hongjun Song and Ping Liu. A novel hybrid color image encryption algorithm using two complex chaotic systems. Optics and Lasers in Engineering 77 (2016): 118–125.

[6] Fowler, A. C., J. D. Gibbon and M. J. McGuinness. The real and complex Lorenz equations and their relevance to physical systems. Physica D: Nonlinear Phenomena 7, no. (1-3) (1983): 126–134.

[7] Gamal M. Mahmoud, Tassos Bountis and Emad E. Mahmoud. Active control and global synchronization of the complex Chen and Lü systems. International Journal of Bifurcation and Chaos 17, no. 12 (2007): 4295–4308.

[8] Emad E. Mahmoud. Generation and suppression of a new hyperchaotic nonlinear model with complex variables. Applied Mathematical Modelling 38, no. 17-18 (2014): 4445–4459.

[9] Chao Luo and Xingyuan Wang. Chaos generated from the fractional-order complex Chen system and its application to digital secure communication. International Journal of Modern Physics C 24, no. 04 (2013): 1350025.

[10] Gamal M. Mahmoud, Emad E. Mahmoud and Mansour E. Ahmed. On the hyperchaotic complex Lü system. Nonlinear Dynamics 58 (2009): 725–738.

[11] Marius-F. Danca, Paul Bourke and Nikolay Kuznetsov. Graphical structure of attraction basins of hidden chaotic attractors: The Rabinovich–Fabrikant system. International Journal of Bifurcation and Chaos 29, no. 01 (2019): 1930001.

[12] Marius-F. Danca, Nikolay Kuznetsov and Guanrong Chen. Unusual dynamics and hidden attractors of the Rabinovich–Fabrikant system. Nonlinear Dynamics 88 (2017): 791–805.

[13] Marius-F. Danca. Hidden transient chaotic attractors of Rabinovich–Fabrikant system. Nonlinear Dynamics 86, no. 2 (2016): 1263–1270.

[14] Marius-F. Danca, Michal Feckan, Nikolay Kuznetsov and Guanrong Chen. Looking more closely at the Rabinovich–Fabrikant system. International Journal of Bifurcation and Chaos 26, no. 02 (2016): 1650038.

[15] Marius-F.Danca. Convergence of a parameter switching algorithm for a class of nonlinear continuous systems and a generalization of Parrondo's paradox. Communications in Nonlinear Science and Numerical Simulation 18, no. 3 (2013): 500–510.

[16] Srivastava, M., S. K. Agrawal, K. Vishal and S. Das. Chaos control of fractional order Rabinovich–Fabrikant system and synchronization between chaotic and chaos controlled fractional order Rabinovich–Fabrikant system. Applied Mathematical Modelling 38, no. 13 (2014): 3361–3372.

[17] Khaled Moaddy, Asad Freihat, Mohammed Al-Smadi, Eman Abuteen and Ishak Hashim. Numerical investigation for handling fractional-order Rabinovich–Fabrikant model using the multistep approach. Soft Computing 22, no. 3 (2018): 773–782.

[18] Sanjay Kumar, Chaman Singh, Sada Nand Prasad, Chandra Shekhar and Rajiv Aggarwal. Synchronization of fractional order Rabinovich-Fabrikant systems using sliding mode control techniques. Archives of Control Sciences 29 (2019).

[19] Yakubu, H. J., E. G. Dada, S. B. Joseph and A. K. Anukem. A new chaotic image encryption algorithm for digital colour images using rabinovich-fabrikant equations. International Journal of Computer Science and Information Security (IJCSIS) 17, no. 1 (2019).

[20] Steven M. Pincus. Approximate entropy as a measure of system complexity. Proceedings of the National Academy of Sciences 88, no. 6 (1991): 2297–2301.

[21] Jan Cermak, Istvan Gyori and Ludek Nechvatal. On explicit stability conditions for a linear fractional difference system. Fractional Calculus and Applied Analysis 18 (2015): 651–672.

[22] Adel Ouannas, Amina Aicha Khennaoui, Shaher Momani, Giuseppe Grassi and Viet-Thanh Pham. Chaos and control of a three-dimensional fractional order discrete-time system with no equilibrium and its synchronization. AIP Advances 10, no. 4 (2020): 045310.

[23] Mohd Taib Shatnawi, Noureddine Djenina, Adel Ouannas, Iqbal M. Batiha and Giuseppe Grassi. Novel convenient conditions for the stability of nonlinear incommensurate fractional-order difference systems. Alexandria Engineering Journal 61, no. 2 (2022): 1655–1663.

[24] Lan-Lan Huang, Ju H. Park, Guo-Cheng Wu and Zhi-Wen Mo. Variable-order fractional discrete-time recurrent neural networks. Journal of Computational and Applied Mathematics 370 (2020): 112633.

[25] Guo–Cheng Wu, De–Qiang Zeng and Dumitru Baleanu. Fractional impulsive differential equations: exact solutions, integral equations and short memory case. Fractional Calculus and Applied Analysis 22, no. 1 (2019): 180–192.

[26] Guo-Cheng Wu, Zhen-Guo Deng, Dumitru Baleanu and De-Qiang Zeng. New variable-order fractional chaotic systems for fast image encryption. Chaos: An Interdisciplinary Journal of Nonlinear Science 29, no. 8 (2019): 083103.

[27] Dumitru Baleanu and Guo–Cheng Wu. Some further results of the laplace transform for variable–order fractional difference equations. Fractional Calculus and Applied Analysis 22, no. 6 (2019): 1641–1654.

[28] Guo-Cheng Wu, Thabet Abdeljawad, Jinliang Liu, Dumitru Baleanu and Kai-Teng Wu. Mittag-Leffler stability analysis of fractional discrete-time neural networks via fixed point technique (2019).

Chapter 7

Dynamical Analysis of Variable Order Discrete-Time Plasma Perturbation Model

This chapter focuses on investigating the chaotic properties of a discrete-time plasma perturbation model with varying order. The study examines the dynamic behavior in three scenarios: when the state variables have commensurate orders, when the orders are non-identical, and when the order varies. The chapter also presents stability results for equilibrium states and the corresponding eigenvalues. Chaotic behavior is illustrated using bifurcation diagrams and the calculation of the Largest Lyapunov exponents. The coexistence of attractors in systems with commensurate orders, along with the accompanying bifurcations, is discussed. A comparison is made between the chaotic behavior of systems with commensurate order, incommensurate order, and variable order, to better understand the advantages of constructing models with variable order. The chapter provides visual support for the state transitions of the system through 3-D attractors. By considering variable orders in model construction, the complex dynamical nature of the plasma perturbation model is revealed.

7.1 Introduction

Matter in nature can exist in various forms and is classified based on its physical properties. Traditionally, matter is categorized as solid, liquid, or gas. However, scientific advancements in the 20th century have revealed the existence of intermediate states that are not observable under normal conditions. One such state is plasma, often referred to as the fourth state of matter. Plasma is an ionized form of gas with unique properties. In plasma, electrons are free to move independently from the atomic nuclei, unlike in the other three states of matter. Plasma is created when gas is superheated, causing the removal of electrons from atoms and resulting in an ionized gas.One fundamental difference between gaseous matter and plasma is how they respond to electrical and magnetic fields. Plasma, being composed of ions, exhibits distinct behavior due to its electrical charge. Plasma can be created by introducing energy to a gas, such as high-voltage electricity, X-rays, gamma rays, or UV rays. Plasma is found in natural sources like the Sun and other stars, as well as in the Earth's ionosphere, where it is formed by UV rays from the Sun. Humans can also create artificial plasma, such as in neon lights, where gas is converted into glowing plasma using electricity.

In 1929, Irving Langmuir, a pioneer in the study of ionized gases, named this state of matter "plasma." The behavior of plasma is influenced by the motion of its particles and the application of external fields. The field of astrophysics has revealed a wide range of natural plasmas in galaxies and interstellar spaces. Many phenomena in space are attributed to interactions between plasma and magnetic particles. Numerous experiments have been conducted to understand the properties of plasma and expand our theoretical knowledge about it. Plasma finds applications in controlling thermonuclear fusion reactions at high temperatures, rocket propulsion utilizing plasma energy, and communication technologies. For instance, the aurora borealis and aurora australis are created by the flow of plasma in the magnetosphere along the Earth's magnetic field. When plasma particles collide with atmospheric gases, it results in the colorful displays.

The study of plasma and its characteristics, particularly in astrophysics, has given rise to the field of plasma physics. This field has attracted the attention of scientists, interdisciplinary researchers, and engineers due to the practical applications of plasma in various domains. Everyday applications of plasma include electric lamps, water purification, medical treatments, energy conversion, lasers, and its role in the composition of the universe with plasma particles. With advancing technology, the exploration of hidden properties of plasma has become more feasible. Investigating the interaction of plasma particles with other materials has helped solve several problems in space science, material and chemical engineering, molecular biology, optical physics, and more. Plasma physics has contributed to astrophysicists' understanding of phenomena such as black holes, the creation and destruction of stars, and the potential existence of life on other planets. Describing the nature of plasma in simple mathematical terms is challenging due to the numerous factors involved in describing its behavior. Theoretical developments in plasma physics incorporate equations from classical mechanics, such as Maxwell's equations, while various mathematical models have been developed to better understand the dynamic nature of plasma particles.

Over the past two decades, there have been numerous research contributions towards modeling plasma and understanding its physical characteristics. Researchers have investigated perturbation models in plasma, such as the work by Constantinescu et al. on plasma perturbation in [9], and the study by Liu et al. on plasma models with resonant magnetic perturbations in [8]. Magnetic perturbations in magnetohydrodynamic plasma models were examined in [12], and a comparison between linear and nonlinear models of plasma in non-axisymmetric perturbations was presented in [13]. Mathematical constructions of plasma models and their applications can be found in [14], while simulation results for plasma models are discussed in the book by Birdsall et al. [15]. Numerical analyses using the homotopy perturbation method were applied in [11], and qualitative analyses of perturbed plasma models were conducted in [10]. The chaotic dynamics of plasma models are discussed in [24–26]. Fractional order modeling, which is a generalization of classical models, has gained attention due to its improved accuracy and consideration of memory effects in systems. Turbulence in plasma models using fractional-type diffusion approaches was explored in [16, 19]. Fractional equation models for plasma dynamics were investigated in [17, 18], and numerical and qualitative analyses of discretized fractional models were discussed in [21]. Fractional models of dusty plasma and cold plasma analysis were presented in [22, 23]. However, no research has been performed considering fractional difference operators in plasma physics. Motivated by the literature and aiming to address this research gap, this chapter focuses on developing a Caputo fractional difference model for plasma perturbation.

7.2 Plasma Perturbation Model (PPM)

In this chapter, we focus on studying the discrete version of a nonlinear 3-D system that describes the dynamic behavior of the displacement of the magnetic field and the pressure gradient of plasma. The system under investigation is discussed in [6].

$$\begin{cases} z_1(\zeta+1) - z_1(\zeta) &= z_2(\zeta)(z_3(\zeta)-1) - \rho_1 z_1(\zeta), \\ z_2(\zeta+1) - z_2(\zeta) &= z_1(\zeta), \\ z_3(\zeta+1) - z_3(\zeta) &= \rho_2(\rho_3 - z_3(\zeta) - (z_2(\zeta))^2 z_3(\zeta)). \end{cases} \tag{7.1}$$

In this study, we aim to investigate the chaotic dynamics of the system described by equation (7.1) with a fractional difference operator [2, 3]. The system incorporates real-valued parameters ρ_1, ρ_2, ρ_3, which represent the relaxation or dissipation of the perturbation, input power to the system, and the relationship between two coefficients of heat diffusion, respectively. Our focus is on exploring the chaotic behavior of this fractional order plasma perturbation model in discrete–time.

$$\begin{cases} \Delta_\kappa^\vartheta z_1(\zeta) &= z_2(\zeta+\vartheta-1)(z_3(\zeta+\vartheta-1)-\rho_1 z_1(\zeta+\vartheta-1), \\ \Delta_\kappa^\vartheta z_2(\zeta) &= z_1(\zeta+\vartheta-1), \\ \Delta_\kappa^\vartheta z_3(\zeta) &= \rho_2(\rho_3 - z_3(\zeta+\vartheta-1)-(z_2(\zeta+\vartheta-1))^2 z_3(\zeta+\vartheta-1)), \end{cases} \tag{7.2}$$

where Δ_κ^ϑ represents the Caputo difference operator of variable order in the range, $0 < \vartheta \leq 1$ with $z_1, z_2, z_3 \in \mathbb{R}^3$ and $\zeta \in \mathbb{N}_{\kappa+1-\vartheta}$. Assuming an initial state for the variables of the system (7.2) as $z_1(0) = \alpha_1$, $z_2(0) = \alpha_2$, and $z_3(0) = \alpha_3$, where α_i are real numbers, we encounter difficulties in directly analyzing the system presented in (7.2). To address this, we employ the method of fractional sum equations proposed in [4] to convert the system into a numerically suitable form. By setting $\kappa = 0$, we obtain the numerical representation as follows:

$$\begin{cases} z_1(\zeta) = & z_1(0) + \sum_{r=1}^{\zeta} \frac{\Gamma(\zeta-r+\vartheta)}{\Gamma(\vartheta)\Gamma(\zeta-r+1)} \left(z_2(r-1)(z_3(r-1)-1)-\rho_1 z_1(r-1)\right), \\ z_2(\zeta) = & z_2(0) + \sum_{r=1}^{\zeta} \frac{\Gamma(\zeta-r+\vartheta)}{\Gamma(\vartheta)\Gamma(\zeta-r+1)} \left(z_1(r-1)\right), \\ z_3(\zeta) = & z_3(0) + \sum_{r=1}^{\zeta} \frac{\Gamma(\zeta-r+\vartheta)}{\Gamma(\vartheta)\Gamma(\zeta-r+1)} \left(\rho_2(\rho_3 - z_3(r-1)-(z_2(r-1))^2 z_3(r-1))\right) \quad \zeta = 1, 2, \cdots. \end{cases} \tag{7.3}$$

The chapter aims at discussing the behavioral aspect of the plasma model under three different scenarios as under.

1. Identical orders for all the state variables.
2. Non-identical real orders for the state variables.
3. Fractional orders as a function of time.

7.3 Commensurate Order PPM Model

In this section, we analyze the stability and chaotic behavior of the system (7.2) when the state variables are assigned identical fractional orders denoted by ϑ. By assigning the same fractional order to all state variables, we treat them as having similar physical characteristics. We explore the chaotic nature of the system (7.2) by generating bifurcation diagrams. These diagrams provide insights into the system's behavior by varying the parameter ρ_2, which represents the input power to the system.

7.3.1 Equilibrium States of (7.2) and their Stability

The analysis of stable states of the fractional difference model (7.2) is performed by obtaining their equilibrium states. The considered system (7.2) has the following three equilibrium states:

1. $ES_1 = (0, 0, \rho_3)$,
2. $ES_2 = \left(0, \sqrt{\rho_3 - 1}, 1\right)$,
3. $ES_3 = \left(0, -\sqrt{\rho_3 - 1}, 1\right)$.

Before performing the analysis, the fundamental theorem necessary for stability analysis of fractional difference systems is given as follows,

Theorem 7.1 *[1] Consider a system of fractional difference equations of order $\vartheta \in (0,1)$ given by*

$$\Delta^{\vartheta}\Psi(n+1-\vartheta) = \Lambda\Psi(n), n = 0, 1, 2, \cdots, \tag{7.4}$$

where $\Lambda \in \mathbb{R}^{k\times k}$. Let

$$\begin{aligned} B^{\vartheta} = &\left\{\delta \in \mathbb{C} : |\delta| < \left(2\cos\left(\frac{|arg\delta| - \pi}{2-\vartheta}\right)\right)^{\vartheta}\right. \\ &\left. and\ |arg\delta| > \frac{\vartheta\pi}{2}\right\}. \end{aligned} \tag{7.5}$$

If all $\delta \in B^{\vartheta}$, then the system (7.4) is asymptotically stable and if $\delta \in \mathbb{C}\backslash cl(B^{\vartheta})$ for any δ then the system (7.4) is unstable.

The analysis will be performed numerically for the following parameters: $\rho_1 = 0.5, \rho_2 = 0.35, \rho_3 = 1.7$ with fractional order $\vartheta = 0.7$. The analysis of the stable state of the system (7.2) is performed based on Theorem 7.1. The equilibrium states for the considered parameter values are $ES_1 = (0, 0, 1.7), ES_{2,3} = (0, \pm 0.8366600, 1)$.

Equilibrium State $ES_1 = (0, 0, 1.7)$:
The Jacobian matrix of the system (7.2) at ES_1 is obtained as

$$V(ES_1) = \begin{bmatrix} -0.5 & 0.7 & 0 \\ 1 & 0 & 0 \\ 0 & 0 & -0.35 \end{bmatrix}. \tag{7.6}$$

Corresponding eigenvalues are $\Lambda_1 = -1.123212460, \Lambda_2 = 0.6232124597, \Lambda_3 = -0.35$. It can be observed that at ES_1, system (7.2) has two negative and one positive eigenvalue. We know that the argument for any positive real number is always zero. From Theorem 7.1, the condition $|arg\Lambda_2| > \frac{\vartheta\pi}{2}$ fails. Thus, the considered system at ES_1 is unstable.

Equilibrium State $ES_2 = (0, 0.8366600265, 1)$:
The Jacobian matrix of the system (7.2) at ES_2 is

$$V(ES_2) = \begin{bmatrix} -0.5 & 0 & 0.8366600265 \\ 1 & 0 & 0 \\ 0 & -0.5856620186 & -0.595 \end{bmatrix}. \tag{7.7}$$

The eigenvalues are $\Lambda_1 = -1.190744513, \Lambda_{2,3} = 0.04787225555 \pm i0.6396995386$. From Theorem (7.1), we get,

$$\left(2\cos\left(\frac{|arg\Lambda_1| - \pi}{2-\vartheta}\right)\right)^{\vartheta} = 1.624504255 > |\Lambda_1| = 1.190744513.$$

$$\left(2\cos\left(\frac{|arg\Lambda_{2,3}| - \pi}{2-\vartheta}\right)\right)^{\vartheta} = 0.6983830782 > |\Lambda_{2,3}| = 0.6414883105.$$

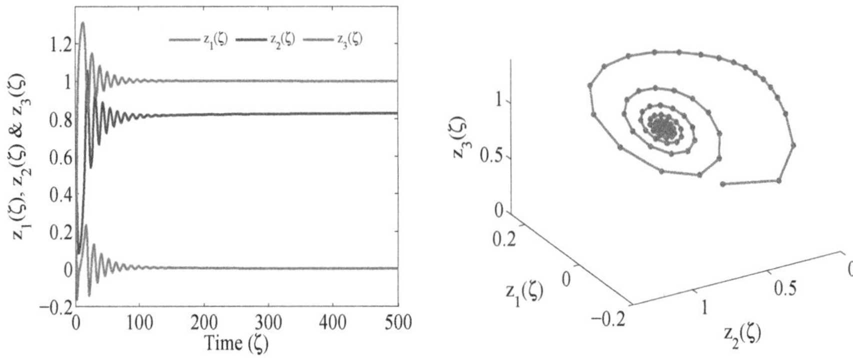

Fig. 7.1: Asymptotic stability of system (7.2) at ES_2.

For the second condition, we find the argument of the eigenvalues,

$$|argument(\Lambda_1)| = 3.141592654 > \frac{\vartheta\pi}{2} = 1.1$$
$$|argument(\Lambda_{2,3})| = 1.496100029 > \frac{\vartheta\pi}{2} = 1.1.$$

Thus, all the eigenvalues of the system (7.2) at ES_2 satisfy the conditions of the stability in Theorem 7.1 ensuring the asymptotic stability of the system.

Equilibrium State $ES_3 = (0, -0.8366600265, 1)$**:**
The Jacobian matrix of the system (7.2) at ES_2 is

$$V(ES_2) = \begin{bmatrix} -0.8 & 0 & -0.8366600265 \\ 1 & 0 & 0 \\ 0 & 0.5856620186 & -0.595 \end{bmatrix}. \tag{7.8}$$

The eigenvalues are $\Lambda_1 = -1.186976377$, $\Lambda_{2,3} = -0.01901180968 \pm i0.5426835287$, which are the same as those obtained at ES_2. Thus, from the previous case, the system (7.2) at ES_3 is asymptotically stable.

The stability analysis results are complemented by numerical simulations, as depicted in Figure 7.1, which includes phase plane portraits and time series plots. The stability of the system is confirmed by observing the convergence of the system's state variables to the equilibrium state $ES_2 = (0, 0.8366600265, 1)$. This is visually represented by an inward moving spiral trajectory observed in the 3-D phase plane plot. These numerical simulations provide further evidence of the system's stability.

7.3.2 Bifurcation Analysis and Chaotic Attractors

Bifurcation diagrams are a useful tool to illustrate the chaotic response of a dynamical system when there is a small change in a parameter value. These diagrams capture the transitions that the system undergoes as the parameter is varied, providing insights into the system's physical characteristics. In this section, we perform behavioral analysis by varying the input parameter (ρ_2) of the system. The other parameters are fixed with values $\rho_1 = 0.5$, $\rho_3 = 1.8$, and $\vartheta = 0.8$. The initial states of the system are set as $(0.1, 0.25, 0.15)$, while ρ_1

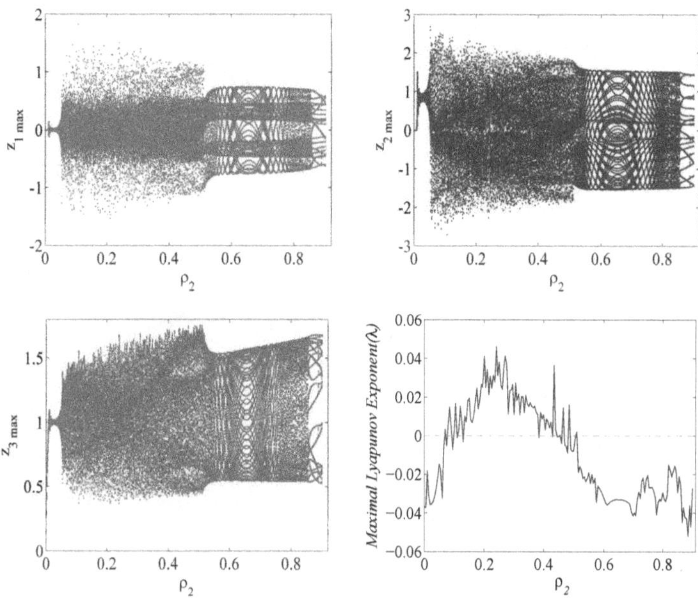

Fig. 7.2: Bifurcation and largest Lyapunov exponents for system (7.2).

is varied in the range $(0, 0.85]$. Bifurcation diagrams are generated for each state variable z_1, z_2, and z_3, and the largest Lyapunov exponents are computed using the Jacobian matrix method proposed by Wu et al. in [5]. The resulting bifurcation diagrams are presented in Figure 7.2.

The bifurcation diagrams shown in Figure 7.2 demonstrate the system's behavior as it transitions from stable states for small values of ρ_2 to a chaotic state. Subsequently, the system enters a non-chaotic state characterized by the formation of limit cycles. The chaotic region is indicated by the largest Lyapunov exponents being above the zero line, while the non-chaotic regions are below the zero line.

To further illustrate the system's transitions between these states, we present 3D phase plane portraits in Figures 7.3 and 7.4. These portraits depict the initial transition from stable spiral trajectories to chaotic attractors for different values of ρ_2, specifically $\rho_2 = 0.03, 0.06, 0.1, 0.25$, as shown in Figure 7.3. Additionally, the reverse transition from the chaotic region to the non-chaotic region can be observed for $\rho_2 = 0.5, 0.75$ in Figure 7.4. These 3D phase plane diagrams provide visual insights into the system's dynamic behavior during the transitions.

7.3.3 Analysis of Coexisting Attractors

The study of coexisting attractors is conducted by varying the initial condition of the state variable z_3. This analysis aims to explore the influence of the initial states of each state variable on the system's characteristics. Furthermore, the parameter value is also varied to examine its significance on the system's dynamics.

Two different initial states are considered for the state variable z_3: $(0.1, 0.25, 0.15)$ and $(0.1, 0.25, -0.15)$. Coexisting bifurcations are obtained for the state variables at these two different initial states, while varying the parameter ρ_1 within the range, $[0.3, 1]$. The other parameter values are set as $\rho_2 = 0.5$, $\rho_3 = 1.8$, and $\vartheta = 0.8$. The resulting coexisting bifurcations, considering the symmetry of initial conditions for the state

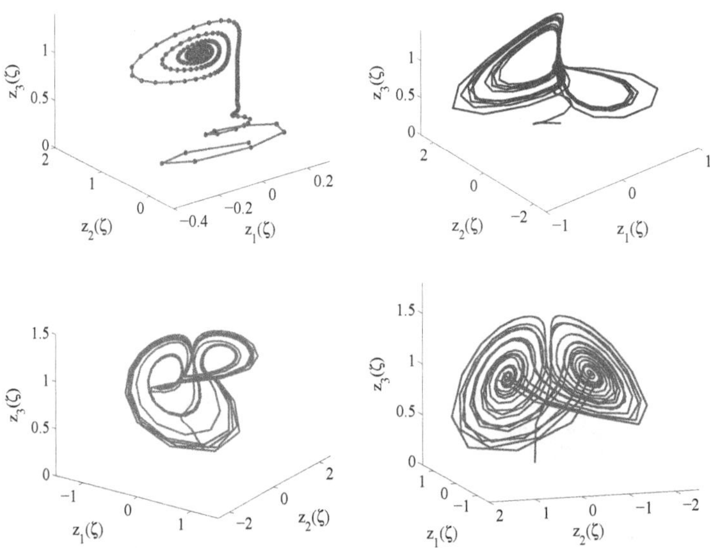

Fig. 7.3: 3-D phase portraits for system (7.2) based on Figure 7.2.

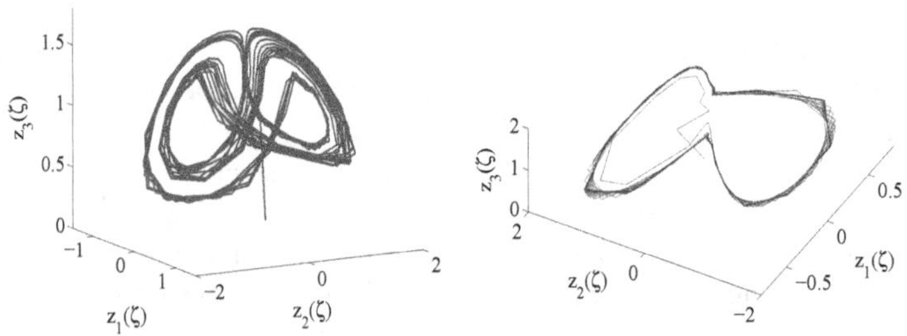

Fig. 7.4: 3-D phase portraits for system (7.2) based on Figure 7.2.

variable z_3, are presented in Figure 7.5. These diagrams provide insights into the coexistence of different attractors based on variations in the initial conditions and the parameter ρ_1.

Upon closer observation of the bifurcation diagrams for each state variable, it becomes evident that the system exhibits an extended chaotic behavior when all initial conditions are on the positive axis. However, for the symmetric initial state $(0.1, 0.25, -0.15)$, the system remains completely unstable throughout this phase. This analysis of the coexistence of attractors provides interesting and informative insights into the complex dynamics of the system.

To further investigate the system dynamics, attractors at four different values of ρ_3 (specifically, $\rho_3 = 0.4, 0.5, 0.65, 0.7$) are illustrated in Figures 7.6, 7.7, 7.8, and 7.9. These attractors are depicted in the form of 2-D plane plots, focusing on each pair of state variables. By analyzing these attractors, one can gain a better understanding of how the system's states change over time.

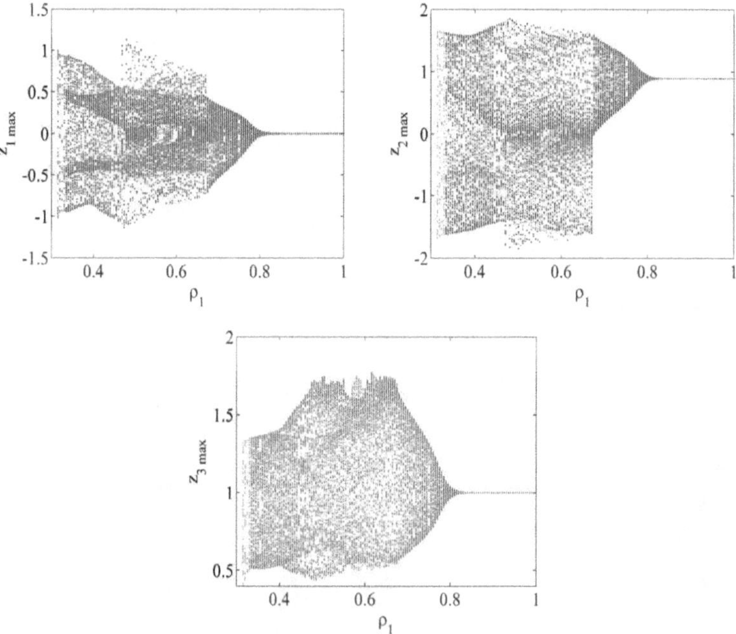

Fig. 7.5: Coexisting bifurcations for system (7.2).

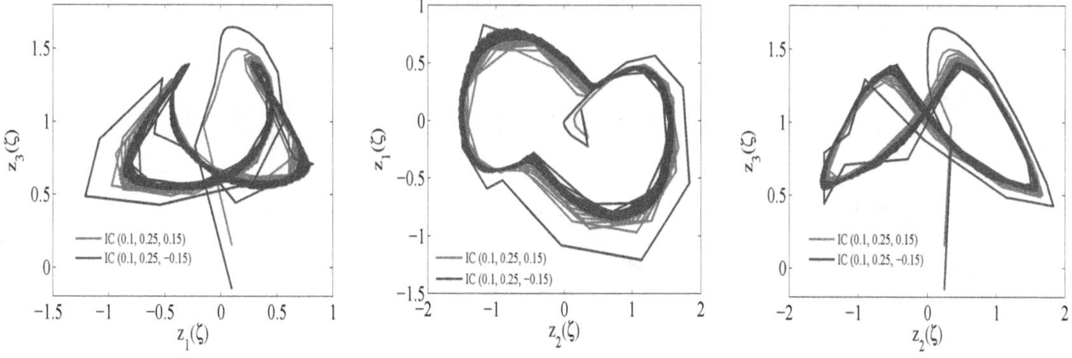

Fig. 7.6: Coexisting phase planes for system (7.2).

7.3.4 Approximate Entropy

The approximate entropy proposed in [20] is a useful tool in representing the randomness of the time series in a dynamical system. The approximate entropy calculations are performed as follows.

$$ApEn(h, r, W) = \Theta_{h+1}(r) - \Theta_h(r), \tag{7.9}$$

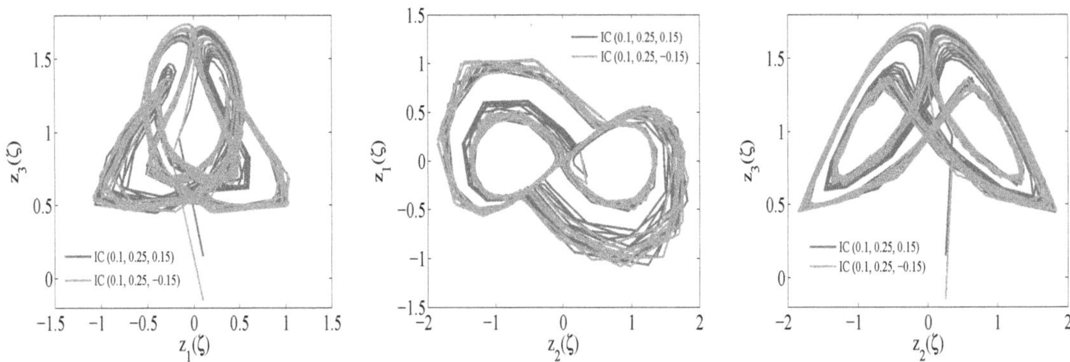

Fig. 7.7: Coexisting phase planes for system (7.2).

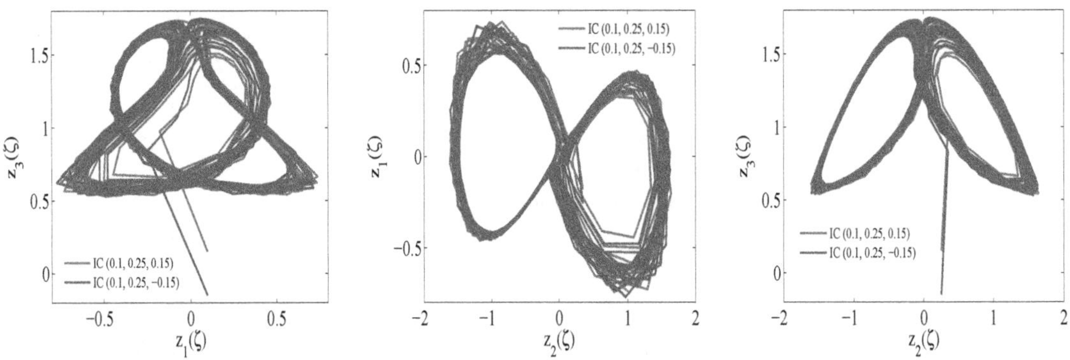

Fig. 7.8: Coexisting phase planes for system (7.2).

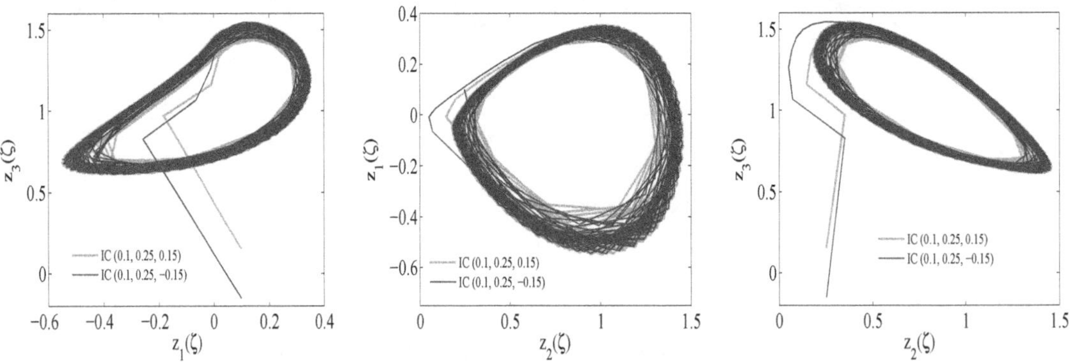

Fig. 7.9: Coexisting phase planes for system (7.2).

where h is the number of data points to be compared, tolerance factor is represented by r and the total time series data is W. Here, $\Theta_h(r) = \dfrac{1}{W-h+1} \sum\limits_{j=1}^{W-h+1} \log M_j^h(r)$, where the value of $M_j^h(r) = \dfrac{|\chi(k) - \chi(l)|}{W-h+1}$,

Table 7.1: Approximate entropy of commensurate order plasma perturbation model.

		$\rho_2 = 0.2$		$\rho_2 = 0.3$		$\rho_2 = 0.4$	
ρ_1	Tolerance (r)	$W = 1500$	$W = 3000$	$W = 1500$	$W = 3000$	$W = 1500$	$W = 3000$
0.5	$r = 0.2$	0.5690	0.5546	0.6916	0.7670	0.7483	0.8461
	$r = 0.3$	0.4831	0.4656	0.5828	0.6555	0.6278	0.6948
	$r = 0.5$	0.3510	0.3252	0.4225	0.4713	0.4840	0.5261
	$r = 0.7$	0.2522	0.2325	0.3190	0.3656	0.3735	0.4085
	$r = 0.9$	0.1839	0.1739	0.2382	0.2858	0.2816	0.3115
0.6	$r = 0.2$	0.5553	0.4933	0.7305	0.6556	0.7809	0.6937
	$r = 0.3$	0.4702	0.4404	0.6217	0.5507	0.6507	0.5748
	$r = 0.5$	0.3274	0.3285	0.4489	0.3976	0.4944	0.4308
	$r = 0.7$	0.2310	0.2320	0.3458	0.2943	0.3815	0.3315
	$r = 0.9$	0.1710	0.1665	0.2663	0.2217	0.2885	0.2460

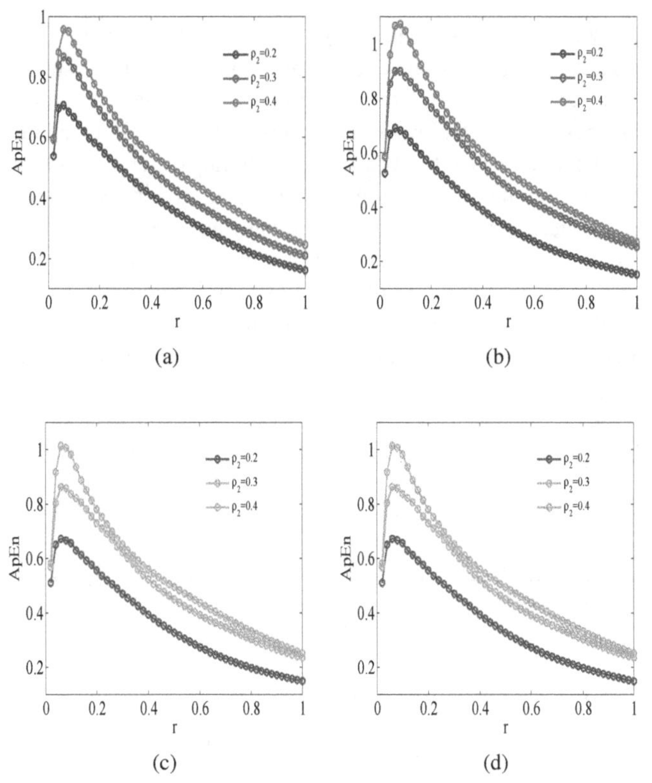

Fig. 7.10: 2-D plot of approximate entropy with varying tolerance (a) $\rho_1 = 0.5$, $W = 1500$, (b) $\rho_1 = 0.5$, $W = 3000$, (c) $\rho_1 = 0.6$, $W = 1500$, (d) $\rho_1 = 0.6$, $W = 3000$.

$\chi(k)$ is the kth block of the time series sequences. The randomness analysis is performed for $\rho_1 = 0.6$, $\rho_3 = 1.8$ and $\vartheta = 0.8$ by varying $\rho_2 = 0.2, 0.3, 0.4$. The values are tabulated in Table 7.1 for different tolerance values and lengths of the chaotic series and plotted in Figure 7.10.

7.4 Incommensurate Order PPM Model

In this section, we explore the behavior of the system with non-identical fractional orders for the state variables. It is important to recognize that each physical system in nature is unique, and the state variables representing these systems exhibit different characteristics. Therefore, understanding the significance of each state variable and its contribution to the system's behavior is crucial.

We investigate the stability and chaotic responses of the system with non-identical fractional orders by varying the parameters. This analysis helps us gain insights into how the system's dynamics are influenced by the individual characteristics of each state variable. By considering the non-identical fractional orders, we can capture the distinct nature of each state variable and its impact on the overall system behavior. Let us consider the discrete–time incommensurate fractional order plasma perturbation model of the form

$$\begin{cases} \Delta_\kappa^{\vartheta_1} z_1(\zeta) &= z_2(\zeta+\vartheta_1-1)(z_3(\zeta+\vartheta_1-1)-1)-\rho_1 z_1(\zeta+\vartheta_1-1), \\ \Delta_\kappa^{\vartheta_2} z_2(\zeta) &= z_1(\zeta+\vartheta_2-1), \\ \Delta_\kappa^{\vartheta_3} z_3(\zeta) &= \rho_2(\rho_3-z_3(\zeta+\vartheta_3-1)-(z_2(\zeta+\vartheta_3-1))^2 z_3(\zeta+\vartheta_3-1)), \end{cases} \tag{7.10}$$

where $\Delta_\kappa^{\vartheta}$ represents the Caputo difference operator of variable order $0 < \vartheta_i \le 1$ for $i = 1,2,3$ with $z_1, z_2, z_3 \in \mathbb{R}^3$ and $\zeta \in \mathbb{N}_{\kappa+1-\vartheta}$. Let the initial state of the variables of the system (7.2) be $z_1(0) = \alpha_1$, $z_2(0) = \alpha_2$ and $z_2(0) = \alpha_3$, here $\alpha_i, i = 1,2,3$ are real numbers. Constructing a model with non-identical orders offers two main advantages. Firstly, it allows us to explore and understand the individual characteristics of the variables that represent the dynamics of the system. By assigning different fractional orders to each state variable, we can investigate their unique contributions to the overall system behavior. Secondly, employing non-identical orders provides numerical advantages by introducing additional degrees of freedom in selecting parameter values. In our case, the introduction of non-identical orders increases the number of parameters to six, offering more flexibility in adjusting and fine-tuning the system's behavior. To facilitate the analysis of the system described in Equation (7.10), we convert it into a numerically suitable form using the method of fractional sum equations proposed in [4]. By setting $\kappa = 0$, we obtain the following numerical form of the system.

$$\begin{cases} z_1(\zeta) = & z_1(0) + \sum\limits_{r=1}^{\zeta} \dfrac{\Gamma(\zeta-r+\vartheta_1)}{\Gamma(\vartheta_1)\Gamma(\zeta-r+1)} \left(z_2(r-1)(z_3(r-1)-1)-\rho_1 z_1(r-1)\right), \\ z_2(\zeta) = & z_2(0) + \sum\limits_{r=1}^{\zeta} \dfrac{\Gamma(\zeta-r+\vartheta_2)}{\Gamma(\vartheta_2)\Gamma(\zeta-r+1)} \left(z_1(r-1)\right), \\ z_3(\zeta) = & z_3(0) + \sum\limits_{r=1}^{\zeta} \dfrac{\Gamma(\zeta-r+\vartheta_3)}{\Gamma(\vartheta_3)\Gamma(\zeta-r+1)} \left(\rho_2(\rho_3-z_3(r-1)-(z_2(r-1))^2 z_3(r-1))\right) \quad \zeta = 1,2,\cdots. \end{cases} \tag{7.11}$$

7.4.1 Analysis of Stability for Equilibrium States

The equilibrium states are investigated for their stable nature numerically considering fixed parametric values for the system (7.10). The conditions required for proving the stability results are obtained from,

Theorem 7.2 *[7] Let* $\alpha_\kappa = \dfrac{\omega_\kappa}{\beta_\kappa} \in (0,1)$, *where* $\kappa = 1,2,3,\cdots,m$, $\omega_\kappa, \beta_\kappa > 0$ *are positive integers, letting the LCM of the denominators to be* Φ. *The trivial solution of the system* (2.5) *with initial condition* $x(0) = x_0$ *is asymptotically stable if the equation*

$$det\left(diag\left(\eta^{\Phi\alpha_1}, \eta^{\Phi\alpha_2}, \ldots, \eta^{\Phi\alpha_m}\right) - (1-\eta^{\Phi} J)\right) = 0, \tag{7.12}$$

has all the roots in the set $\mathbb{C}\backslash B_1^{\vartheta}$, *where* $\vartheta = \frac{1}{\Phi}$ *and*

$$B_1^{\vartheta} = \left\{ \eta \in \mathbb{C} : |\eta| \leq \left(2\cos\frac{|arg\eta|}{\vartheta}\right)^{\vartheta} \text{ and } |arg\eta| \leq \frac{\vartheta\pi}{2} \right\}.$$

The investigation of the stability properties of the incommensurate order system (7.10) are performed for the parameters assuming $\rho_1 = 0.55$, $\rho_2 = 0.15$, $\rho_3 = 1.8$, $\vartheta_1 = 0.9$, $\vartheta_1 = 0.6$, $\vartheta_3 = 0.8$ with initial states of the system at $(0.1, 0.25, 0.15)$. The equilibrium states of the system for the considered values of the parameters are,

1. $ES_1 = (0, 0, 1.8)$,
2. $ES_2 = (0, 0.8944271910, 1)$,
3. $ES_3 = (0, -0.8944271910, 1)$.

We now proceed to analyse the stability characteristics based on the conditions in Theorem (7.2).

Equilibrium State $ES_1 = (0, 0, 1.8)$**:** The stability of the incommensurate order system (7.10) is investigated with eigenvalues obtained from the Jacobian matrix evaluated at the equilibrium point ES_1. Let the matrix based on equation 7.12 be denoted as

$$\Omega = diag\left(\eta^{\Phi\alpha_1}, \eta^{\Phi\alpha_2}, \ldots, \eta^{\Phi\alpha_m}\right) - (1 - \eta^{\Phi}J).$$

Matrix Ω can be obtained as

$$\Omega = \begin{bmatrix} \eta^9 - 0.55\eta^{10} + 0.55 & -0.8 + 0.8\eta^{10} & 0 \\ \eta^{10} - 1 & \eta^6 & 0 \\ 0 & 0 & \eta^8 - 0.15\eta^{10} + 0.15 \end{bmatrix}. \tag{7.13}$$

Determinant of (7.13) yields

$$\begin{aligned} &\eta^{23} - 0.15\eta^{25} + 0.15\eta^{15} - 0.55\eta^{24} + 0.0825\eta^{26} - 0.1650\eta^{16} + 0.55\eta^{14} \\ &+ 0.0825\eta^6 + 1.6\eta^{18} - 0.360\eta^{20} + 0.360\eta^{10} - 0.8\eta^8 - 0.120 - 0.8\eta^{28} + 0.120\eta^{30} = 0. \end{aligned} \tag{7.14}$$

Roots of the equation (7.14) are given in Table 7.2 From numerical calculations based on Theorem 7.2, it can be observed that the eigenvalue η_{11} satisfies the conditions $|\eta_{11}| < (2\cos(10\,|arg\eta_{11}|))^{0.1}$ and $|arg\eta_{11}| < \frac{\pi}{20}$ implying the presence of η_{11} in the set B_1^{ϑ}. Thus, the system (7.10) is not stable at ES_1 for the considered parameters value.

Equilibrium State $ES_2 = (0, 0.8944271910, 1)$**:** Let the matrix based on equation 7.12 be denoted as

$$\Omega = diag\left(\eta^{\Phi\alpha_1}, \eta^{\Phi\alpha_2}, \ldots, \eta^{\Phi\alpha_m}\right) - (1 - \eta^{\Phi}J).$$

Matrix Ω can be obtained as

$$\Omega = \begin{bmatrix} \eta^9 - 0.55\eta^{10} + 0.55 & 0 & -0.8944271910(-1 + \eta^{10}) \\ \eta^{10} - 1 & \eta^6 & 0 \\ 0 & 0.2683281573(1 - \eta^{10}) & \eta^8 - 0.27\eta^{10} + 0.27 \end{bmatrix}. \tag{7.15}$$

Table 7.2: Roots of the equation (7.14).

η_1 = -2.582086901	η_{16} = 0.3033116103+0.8602260942 i
η_2 = 2.582086901	η_{17}= -0.2301315796+0.9465602288 i
η_3 = -0.7320115501-0.3110643607 i	η_{18} = -0.4044872452+0.8521404319 i
η_4 = 0.7320115501+0.3110643607 i	η_{19} = -0.9533728281+0.7363934572 i
η_5 = -0.2940021646-0.7251001268 i	η_{20}= -0.7491577603+0.5568161476 i
η_6 = 0.2940021646+0.7251001268 i	η_{21} = -0.9406212103+0.08050719092 i
η_7 = -0.2940021646+0.7251001268 i	η_{22} = -0.9406212103-0.08050719092 i
η_8 = 0.2940021646-0.7251001268 i	η_{23} = -0.7491577603-0.5568161476 i
η_9 = -0.7320115501+0.3110643607 i	η_{24} = -0.9533728281-0.7363934572 i
η_{10} = 0.7320115501-0.3110643607 i	η_{25} = -0.4044872452-0.8521404319 i
η_{11} = 0.8923248110	η_{26} = -0.2301315796-0.9465602288 i
η_{12} = 1.113724617	η_{27}= 0.3033116103-0.8602260942 i
η_{13} = 0.8588604732+0.5094540037 i	η_{28} = 0.3959523154-1.067460425 i
η_{14} = 0.7166215104+0.6854546968 i	η_{29} = 0.7166215104-0.6854546968 i
η_{15} = 0.3959523154+1.067460425 i	η_{30}= 0.8588604732-0.5094540037 i

Determinant of (7.15) yields,

$$\begin{aligned}&\eta^{23} - 0.27\eta^{25} + 0.27\eta^{15} - 0.55\eta^{24} + 0.1485\eta^{26} - 0.2970\eta^{16} + 0.55\eta^{14}\\&+ 0.1485\eta^{6} - 0.720\eta^{10} + 0.72\eta^{20} + 0.24 - 0.24\eta^{30} = 0.\end{aligned} \tag{7.16}$$

Roots of the equation (7.16) are given in Table 7.3 It is clear that $|\eta_l| > (2\cos(10\,|arg\eta_l|))^{0.1}$ for $l \in \{1,2,3,5,6,9,14,18,23,26,27,29,30\}$ and $|arg\eta_l| > \frac{\pi}{20}$ for $l \in \{3,4,5,6,7,8,9,10,11,12,13,14,15,16,17,18,19,20,21,22,23,24,25,26,27,28,29\}$. From Theorem 7.1, $\eta_l \in \mathbb{C}\backslash B_1^{\vartheta}$ for $1 \le l \le 30$, ensuring the asymptotic stability of the system (7.10) at equilibrium point ES_2.

Table 7.3: Roots of the equation (7.16).

η_1 = 1.168239224	η_{16} = -0.9152197686
η_2 = 0.9143375748+0.1284562573 i	η_{17}= -0.9658249279-0.2217725494 i
η_3 = 0.8711341767+0.4841454622 i	η_{18} = -1.131941174-0.5473935919 i
η_4 = 0.6851839915+0.5165141604 i	η_{19} = -0.7400880754-0.4995215296 i
η_5 = 0.9314839905+0.7973748935 i	η_{20}= -0.6967912035-0.6023363212 i
η_6 = 0.4419989860+0.9137706200 i	η_{21} = -0.3531699687-0.8362785147 i
η_7 = 0.3133956304+0.8499399424 i	η_{22} = -0.2318808674-0.9004230428 i
η_8 = 0.07106790376+0.9260292086 i	η_{23} = -0.2354157647-1.331715135 i
η_9 = -0.2354157647+1.331715135 i	η_{24} = 0.07106790376-0.9260292086 i
η_{10} = -.2318808674+.9004230428 i	η_{25} = 0.3133956304-.8499399424 i
η_{11} = -0.3531699687+0.8362785147 i	η_{26} = 0.4419989860-0.9137706200 i
η_{12} = -0.6967912035+0.6023363212 i	η_{27}= 0.9314839905-0.7973748935 i
η_{13} = -0.7400880754+0.4995215296 i	η_{28} = 0.6851839915-0.5165141604 i
η_{14} = -1.131941174+0.5473935919 i	η_{29} = 0.8711341767-0.4841454622 i
η_{15} = -0.9658249279+0.2217725494 i	η_{30}= 0.9143375748-0.1284562573 i

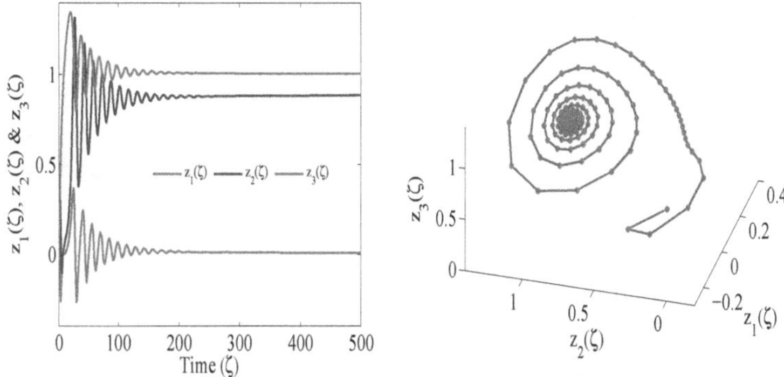

Fig. 7.11: Asymptotic Stability of System (7.10) at ES_2.

Equilibrium State ES_3**:** Let the matrix based on equation 7.12 be denoted as

$$\Omega = diag\left(\eta^{\Phi\alpha_1}, \eta^{\Phi\alpha_2}, \ldots, \eta^{\Phi\alpha_m}\right) - (1 - \eta^{\Phi} J).$$

Matrix Ω can be obtained as,

$$\Omega = \begin{bmatrix} \eta^9 - 0.55\eta^{10} + 0.55 & 0 & -0.8944271910(-1 - \eta^{10}) \\ \eta^{10} - 1 & \eta^6 & 0 \\ 0 & -0.2683281573(-1 - \eta^{10}) & \eta^8 - 0.27\eta^{10} + 0.27 \end{bmatrix}. \tag{7.17}$$

Determinant of (7.17) yields,

$$\begin{aligned} &\eta^{23} - 0.27\eta^{25} + 0.27\eta^{15} - 0.55\eta^{24} + 0.1485\eta^{26} - 0.297\eta^{16} + 0.55\eta^{14} \\ &+ 0.1485\eta^6 - 0.720\eta^{10} + 0.72\eta^{20} + 0.24 - 0.24\eta^{30} = 0. \end{aligned} \tag{7.18}$$

The obtained roots of the equation (7.18) are the same the values given in Table 7.3. Therefore, it is evident that the system (7.10) is asymptotically stable at ES_3. Numerical conditions for the stability of the equilibrium states ES_1, ES_2, and ES_3 in the system (7.10) have been derived. Similar to the commensurate order system (7.2), the system (7.10) exhibits stability only at equilibrium states ES_1 and ES_2, while it is unstable at ES_0. The stable behavior of the system, considering the assumed parameter values, is demonstrated through time-varying plots and 3D phase plane trajectories in Figure 7.11. Initially, the state variables experience oscillations for a short period, but as time progresses, the trajectories converge towards the equilibrium state ES_2. This convergence is supported by inward-moving spiral trajectories originating from the initial state of the system and approaching ES_2.

7.4.2 Chaotic Response of the Non-Identical Order PPM Model

This section focuses on investigating the dynamic changes that occur in the system (7.10) when non-identical orders are introduced to the state variables. To understand the significant variations caused by non-identical orders, a study is conducted using the same set of numerical values as in the commensurate order system.

For the numerical simulations, the parameter values are chosen as $\rho_1 = 0.5$, $\rho_3 = 1.8$, and the fractional orders are set to $\vartheta_1 = 0.8$, $\vartheta_2 = 0.75$, $\vartheta_3 = 0.9$. The initial states are given as $(0.1, 0.25, 0.15)$. Bifurcation

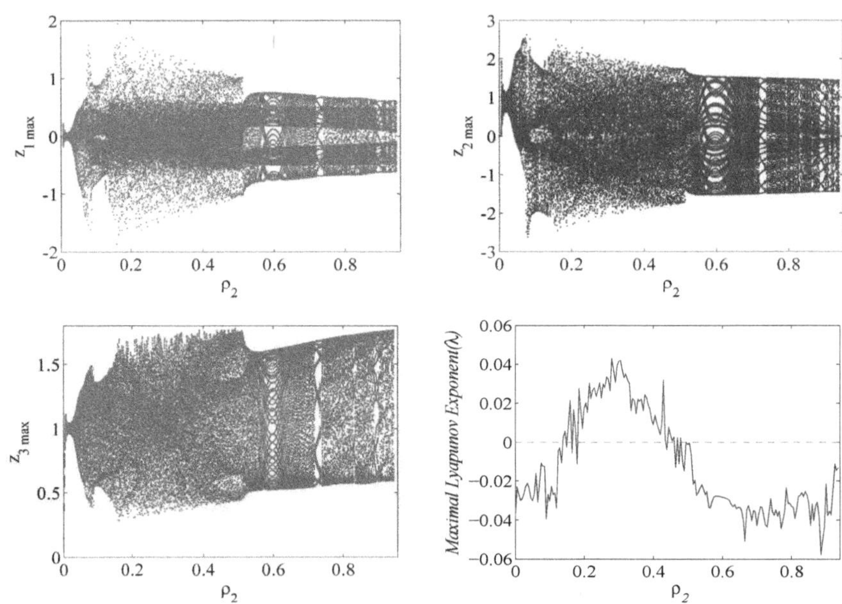

Fig. 7.12: Bifurcation and largest Lyapunov exponents for system (7.10).

diagrams and the largest Lyapunov exponent are presented in Figure 7.12, with the parameter ρ_2 varying within the range $[0, 0.94]$. The breakdown at different values of ρ_2 is illustrated through phase plane plots in Figures 7.13 and 7.14.

A comparative analysis is conducted between the bifurcation diagrams obtained for the commensurate order system (7.2) and the non-identical order system (7.10). Upon visual observation, the bifurcation diagrams clearly exhibit an extended region of bifurcation in the case of the non-identical order system, which is supported by the largest Lyapunov exponents. The region of chaotic response in both cases is similar, but the formation of periodic behavior after the chaotic region is more pronounced in the non-identical order system.

When the analysis is further examined based on the attractors at different values of the parameter ρ_2, it can be observed that in the commensurate order system, the system transitions directly from the stable state to the chaotic state, as shown in Figures 7.3 and 7.4. However, in the non-identical order system, an intermediate periodic state emerges during the transition from the stable state to the chaotic state, as illustrated in the 3-D plots of Figure 7.13. This extended periodic behavior is an additional characteristic that arises when the system is constructed with different fractional orders for each state variable. In this particular study, the fractional order values are very close to 0.8 for state variable z_1, while z_2 and z_3 have orders of 0.75 and 0.9, respectively.

7.4.3 Impact of Fractional Order on System Dynamics

This section of the chapter focuses on investigating the impact of the fractional order of the state variable z_2 at $\vartheta_2 = 0.45, 0.7, 0.85$ on the overall system dynamics, while varying the parameter $\rho_2 \in [0, 1]$. The remaining parameters are set to $\rho_1 = 0.5$, $\rho_3 = 1.8$, $\vartheta_1 = 0.8$, $\vartheta_3 = 0.9$, and the initial states are $(0.1, 0.25, 0.15)$. The simultaneous bifurcations at three different values of ϑ_2 are presented in Figure 7.15, and the impact is

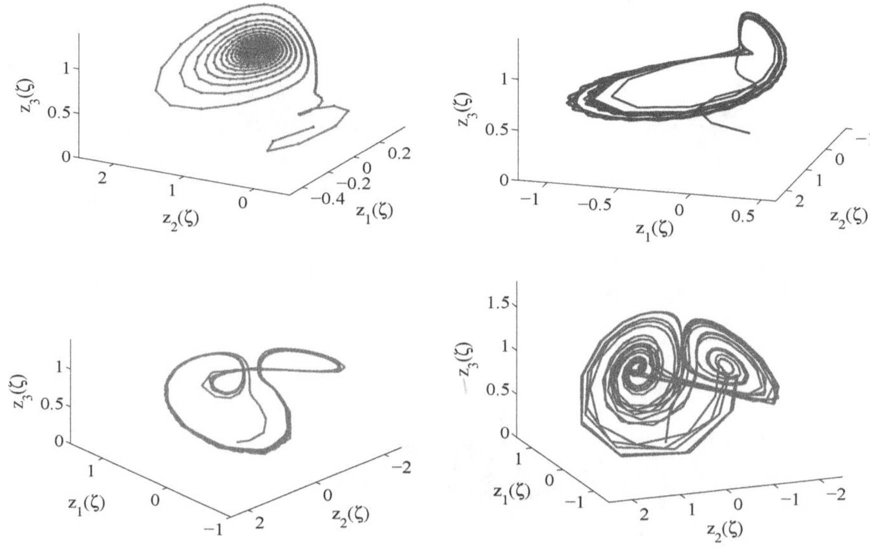

Fig. 7.13: 3-D phase portraits for system (7.10) based on Figure 7.2.

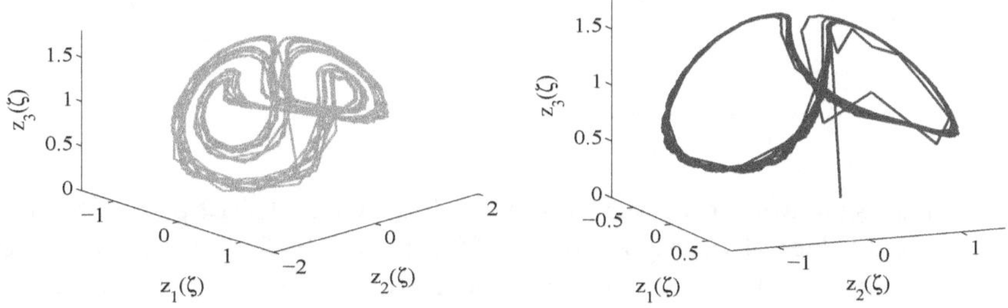

Fig. 7.14: 3-D phase portraits for system (7.10) based on Figure 7.2.

visualized using 3-D phase plane portraits in Figure 7.16. The bifurcation diagrams illustrate the simultaneous effect of the fractional order ϑ_2 and the parameter ρ_2 through a series of plots. The observations from these diagrams provide insights into the relationship between the fractional order of the system and its chaotic behavior.

It is observed that when the fractional order of the system is lower, the system does not exhibit chaotic behavior compared to higher fractional values. For instance, when $\vartheta_2 = 0.45$, the system demonstrates stable behavior over a wide range of ρ_2 values compared to the cases when $\vartheta_2 = 0.7$ and $\vartheta_2 = 0.85$. Although the periodic oscillations of the system coincide for the considered three fractional orders at certain values of ρ_2, the difference lies in the transition behavior. For $\vartheta_2 = 0.45$, the system directly shifts from a stable state to a periodic state, while for $\vartheta_2 = 0.7$ and $\vartheta_2 = 0.85$, an intermediate chaotic state is observed before reaching the periodic state. This transition in the system's behavior provides valuable insights into the physical characteristics of the plasma system under different circumstances.

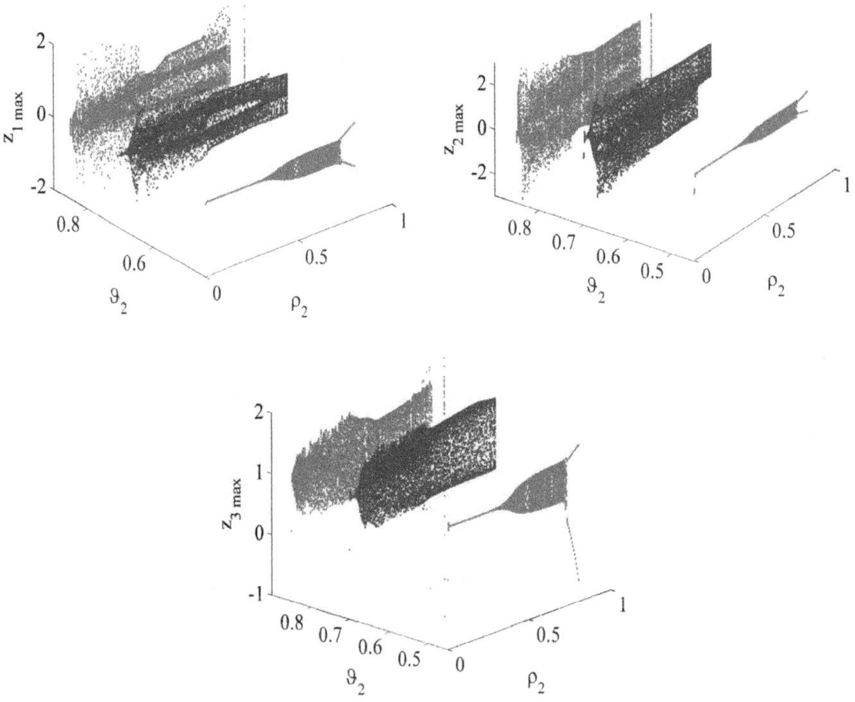

Fig. 7.15: Bifurcation for system (7.10).

To further understand the significant changes in the system's behavior, 3-D phase plane plots are presented in Figure 7.16 for two fixed values of ρ_2, namely $\rho_2 = 0.25$ and $\rho_2 = 0.65$. These plots demonstrate the distinct changes in the system's behavior as influenced by the fractional order of the state variable.

7.4.4 Randomness of Chaotic Time Series of Incommensurate Order Model

Randomness of the chaotic series of incommensurate order plasma model is illustrated with approximate entropy calculated based on (7.9). The complexity analysis is performed for $\rho_1 = 0.6$, $\rho_3 = 1.8$ and $\vartheta_1 = 0.8$, $\vartheta_2 = 0.75$, $\vartheta_3 = 0.9$ by varying $\rho_2 = 0.2, 0.3, 0.4$. The values are tabulated in Table 7.4 for different tolerance values and lengths of the chaotic series and visualized in Figure 7.17.

7.5 Dynamics of the Plasma Perturbation Model for Variable Order

In this section of the chapter, we extend the discussion from the previous section by considering the concept of non-identical orders for the state variables in a time-varying manner. It is important to study the nature of the state variables while preserving their specific physical characteristics. Since nothing in nature remains constant and everything evolves over time, introducing the time-varying nature of the state variables allows for a more accurate representation of the behavioral aspects of real-world systems.

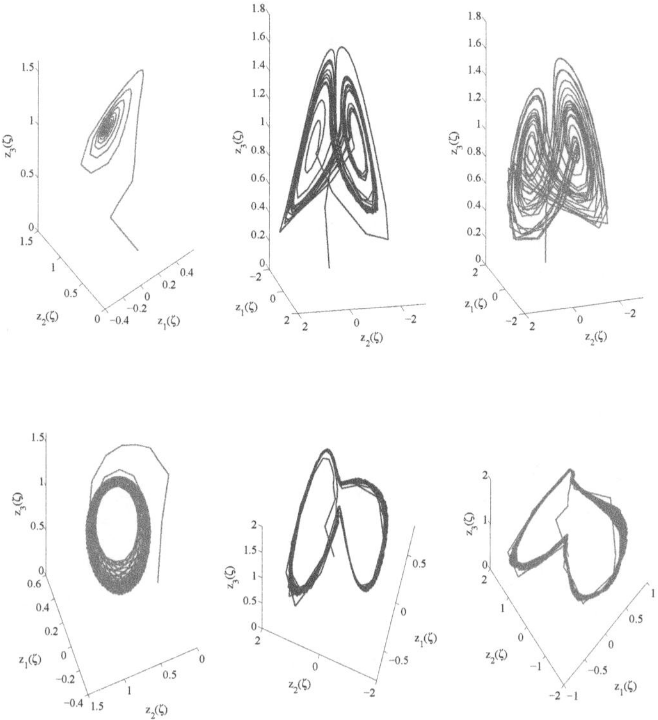

Fig. 7.16: 3-D phase portraits for system (7.10) based on Figure 7.15.

Table 7.4: Approximate entropy of incommensurate order plasma perturbation model.

		$\rho_2 = 0.2$		$\rho_2 = 0.3$		$\rho_2 = 0.4$	
ρ_1	Tolerance (r)	$W = 1500$	$W = 3000$	$W = 1500$	$W = 3000$	$W = 1500$	$W = 3000$
	$r = 0.2$	0.5616	0.5877	0.6442	0.5857	0.6862	0.6648
	$r = 0.3$	0.4857	0.4994	0.5335	0.4882	0.5737	0.5617
0.5	$r = 0.5$	0.3578	0.3656	0.3724	0.3322	0.4388	0.4409
	$r = 0.7$	0.2596	0.2742	0.2754	0.2472	0.3371	0.3367
	$r = 0.9$	0.1952	0.2106	0.2091	0.1859	0.2469	0.2443
	$r = 0.2$	0.5777	0.5201	0.6077	0.5524	0.6178	0.5058
	$r = 0.3$	0.4949	0.4701	0.5351	0.4852	0.5352	0.4687
0.6	$r = 0.5$	0.3628	0.3593	0.3836	0.3917	0.4279	0.3974
	$r = 0.7$	0.2683	0.2622	0.2800	0.2958	0.3246	0.3063
	$r = 0.9$	0.2003	0.1954	0.2016	0.2152	0.2367	0.2295

The concept of variable order discrete-time systems has been relatively underexplored in research, with only a few studies addressing the topic. For example, the chaotic dynamics study of the Tinkerbell map in [27] explores the variable order nature of the system. Additionally, the application of variable order models with fractional difference operators in neural networks is discussed in [28], while the application towards

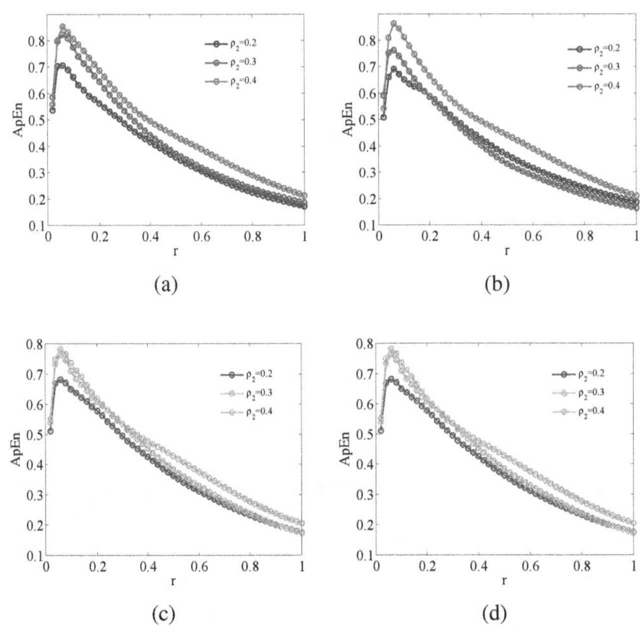

Fig. 7.17: 2-D plot of approximate entropy with varying tolerance (a) $\rho_1 = 0.5$, $W = 1500$, (b) $\rho_1 = 0.5$, $W = 3000$, (c) $\rho_1 = 0.6$, $W = 1500$, (d) $\rho_1 = 0.6$, $W = 3000$.

secure communications is presented in [29]. These studies highlight the potential benefits and applications of variable order models in various domains.

To achieve a higher level of accuracy in modeling the system's behavior, we consider the time-varying fractional orders for the state variables. The orders of the state variables are assumed to be functions of a time variable ζ, and the proposed variable order plasma perturbation model takes the following form:

$$\begin{cases} \Delta_{\kappa}^{\vartheta(\zeta)} z_1(\zeta) &= z_2(\zeta+\vartheta(\zeta)-1)(z_3(\zeta+\vartheta(\zeta)-1)-1)-\rho_1 z_1(\zeta+\vartheta(\zeta)-1), \\ \Delta_{\kappa}^{\vartheta(\zeta)} z_2(\zeta) &= z_1(\zeta+\vartheta(\zeta)-1), \\ \Delta_{\kappa}^{\vartheta(\zeta)} z_3(\zeta) &= \rho_2(\rho_3 - z_3(\zeta+\vartheta(\zeta)-1)-(z_2(\zeta+\vartheta(\zeta)-1))^2 z_3(\zeta+\vartheta(\zeta)-1)), \end{cases} \tag{7.19}$$

where $\Delta_{\kappa}^{\vartheta}$ represents the Caputo difference operator of variable orders $0 < \vartheta(\zeta) \leq 1$. Let the initial state of the variables of the system (7.19) be $z_1(0) = \alpha_1$, $z_2(0) = \alpha_2$ and $z_2(0) = \alpha_3$, here $\alpha_i, i = 1, 2, 3$ are real numbers. Using the method of fractional sum equations proposed in [22], the proposed model (7.19) is converted to a numerically suitable model. By letting $\kappa = 0$, the numerical form is obtained as

$$\begin{cases} z_1(\zeta) = & z_1(0) + \sum\limits_{r=1}^{\zeta} \dfrac{\Gamma(\zeta - r + \vartheta(r))}{\Gamma(\vartheta(r))\Gamma(\zeta - r + 1)} \left(z_2(r-1)(z_3(r-1)-1) - \rho_1 z_1(r-1)\right), \\ z_2(\zeta) = & z_2(0) + \sum\limits_{r=1}^{\zeta} \dfrac{\Gamma(\zeta - r + \vartheta(r))}{\Gamma(\vartheta(r))\Gamma(\zeta - r + 1)} \left(z_1(r-1)\right), \\ z_3(\zeta) = & z_3(0) + \sum\limits_{r=1}^{\zeta} \dfrac{\Gamma(\zeta - r + \vartheta(r))}{\Gamma(\vartheta(r))\Gamma(\zeta - r + 1)} \left(\rho_2(\rho_3 - z_3(r-1) - (z_2(r-1))^2 z_3(r-1))\right) \quad \zeta = 1, 2, \cdots. \end{cases} \tag{7.20}$$

Here we perform the study based on two different functions of $\vartheta(\zeta)$ considered as follows:

Case 1: $\vartheta(\zeta) = 0.1\sin(\zeta) + 0.75$.
Case 2: $\vartheta(\zeta) = 0.2\sin\left(\frac{1}{0.01+\zeta}\right) + 0.75$.

7.5.1 Chaos Analysis for $\vartheta(\zeta) = 0.1\sin(\zeta) + 0.75$

In this section, we examine the impact of introducing variable orders, represented by a function of the time variable ζ, in the plasma perturbation model. The parameter values used for the simulations are the same as those in Section 7.3, with the only difference being that the fractional order is now a function of ζ given by $\vartheta(\zeta) = 0.1\sin(\zeta) + 0.75$. The parameter values are $\rho_1 = 0.5$, $\rho_3 = 1.8$, and the initial states are $(0.1, 0.25, 0.15)$, while ρ_2 varies in the range of $[0, 0.85]$. To analyze the chaotic dynamics of the system (7.19), we present bifurcation diagrams that illustrate the system's behavior, and the regions of chaos are identified using the largest Lyapunov exponents, as shown in Figure 7.18. Additionally, the breakdown of the bifurcation diagrams at different values of $\rho_1 = 0.1, 0.3, 0.55, 0.75$ is presented in Figure 7.19, which explains the transition from chaotic behavior to the convergence of the system to a stable state.

By introducing variable orders in the plasma perturbation model, we can observe the influence of time on the system's behavior and its impact on the chaotic dynamics. The varying fractional orders capture the time-dependent nature of the system, resulting in unique patterns and transitions in the bifurcation diagrams, offering insights into the complex dynamics of the plasma system.

By introducing variable orders for the state variables, the model exhibits a more complex and diverse dynamical behavior compared to the commensurate order system (7.2). A comparison between Figure 7.18 and Figure 7.2 reveals two distinct characteristics. Firstly, the range of stability for the system with variable orders (Case 1) is larger than that of the constant commensurate order system. This means that the system

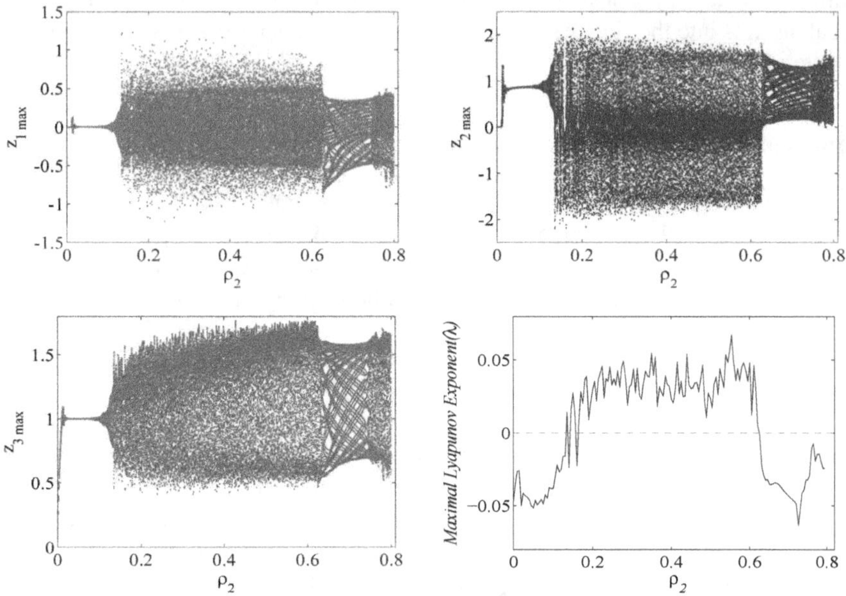

Fig. 7.18: Bifurcation and largest Lyapunov exponents for system (7.19) with variable order $\vartheta(\zeta) = 0.1\sin(\zeta) + 0.75$.

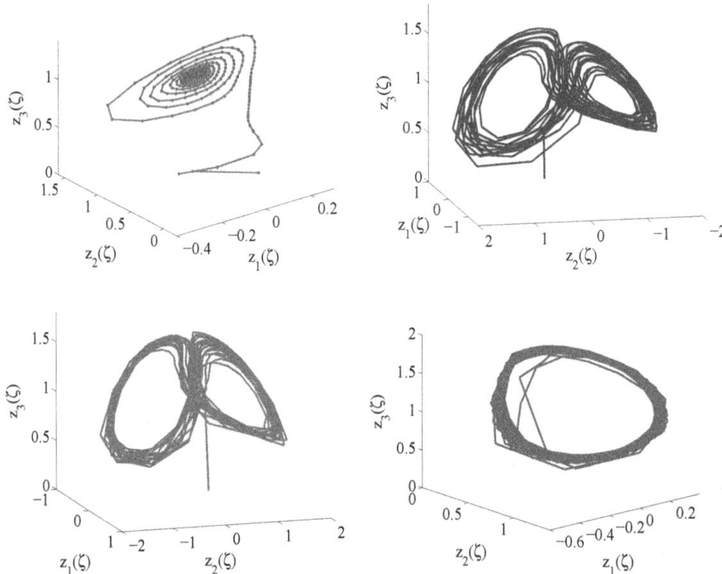

Fig. 7.19: 3-D phase portraits for system (7.19) based on Figure 7.18 with variable order $\vartheta(\zeta) = 0.1\sin(\zeta) + 0.75$.

can remain in a stable state over a wider range of parameter values when variable orders are considered. Secondly, the range of chaotic behavior increases while the region of periodic oscillation is reduced in the variable order system. This indicates that the system exhibits more chaotic dynamics and fewer periodic oscillations when the orders of the state variables vary with time. To further analyze the shifts in states, phase plane plots for different values of ρ_2 (namely, $\rho_2 = 0.1, 0.3, 0.55, 0.75$) are presented in Figure 7.19. These plots provide visual insights into the changes in the system's behavior, showcasing the transition from stable states to chaotic states as the parameter values vary.

7.5.2 Randomness of Variable Order Chaotic Series

Randomness of the chaotic series of incommensurate order plasma model is illustrated with approximate entropy calculated based on (7.9). The randomness analysis is performed for $\rho_1 = 0.6$, $\rho_3 = 1.8$ and $\vartheta = \vartheta(\zeta) = 0.1\sin(\zeta) + 0.75$ by varying $\rho_2 = 0.2, 0.3, 0.4$. The values are tabulated in Table 7.5 for different tolerance values and lengths of the chaotic series with simulations in Figure 7.20.

7.5.3 Chaos Analysis for $\vartheta(\zeta) = 0.2\sin\left(\frac{1}{0.01+\zeta}\right) + 0.75$

In this section, the focus is on studying the chaotic response of the system for varying values of the parameter ρ_2, while keeping $\rho_1 = 0.5$ and $\rho_3 = 1.8$ constant. The initial states are set as $(0.1, 0.25, 0.15)$. The chaotic behavior of the system is analyzed through the use of bifurcation diagrams and the calculation of the largest Lyapunov exponents. Figure 7.21 presents the bifurcation diagrams, highlighting the regions of chaotic response as the parameter ρ_2 varies within the range of [0, 1]. The largest Lyapunov exponents provide quantitative information about the system's sensitivity to initial conditions and the presence of chaotic

Table 7.5: Approximate entropy of variable order plasma perturbation model.

		$\vartheta(\zeta) = 0.1\sin(\zeta) + 0.75$		
ρ_1	Tolerance (r)	$\rho_2 = 0.2$	$\rho_2 = 0.3$	$\rho_2 = 0.4$
	$r = 0.2$	0.5138	0.5894	0.5646
	$r = 0.3$	0.4521	0.5064	0.4914
0.5	$r = 0.5$	0.3362	0.3769	0.4064
	$r = 0.7$	0.2398	0.2765	0.3129
	$r = 0.9$	0.1774	0.1993	0.2297
	$r = 0.2$	0.4067	0.5142	0.5906
	$r = 0.3$	0.3684	0.4324	0.5226
0.6	$r = 0.5$	0.2704	0.2879	0.3892
	$r = 0.7$	0.1921	0.2054	0.2840
	$r = 0.9$	0.1393	0.1528	0.2046

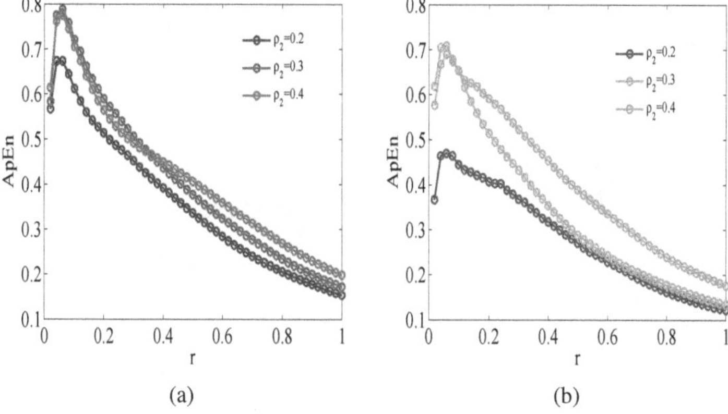

Fig. 7.20: 2-D plot of approximate entropy with varying tolerance (a) $\rho_1 = 0.5$, $W = 1500$, (b) $\rho_1 = 0.6$, $W = 1500$.

dynamics. Furthermore, the behavioral analysis of the state variables is conducted at different values of ρ_2, specifically at $\rho_2 = 0.125, 0.25, 0.65, 0.825$. This analysis is visualized through phase portraits, which depict the trajectories and patterns exhibited by the state variables. The phase portraits are presented in Figure 7.22. By examining the bifurcation diagrams, Lyapunov exponents, and phase portraits, a comprehensive understanding of the chaotic response and system dynamics for varying values of ρ_2 can be gained. The system, characterized by the variable order $\vartheta(\zeta) = 0.2\sin\left(\frac{1}{0.01+\zeta}\right) + 0.75$, exhibits a wide range of chaotic behaviors as the parameter ρ_2 varies. Notably, the system demonstrates the presence of a periodic window, which is evident from the bifurcation diagrams. This periodic window is supported by the negative values of the largest Lyapunov exponents, as shown in Figure 7.21. Therefore, it is evident that incorporating variable orders into the system allows for capturing the distinct characteristics and behaviors of the system, including the emergence of periodicity.

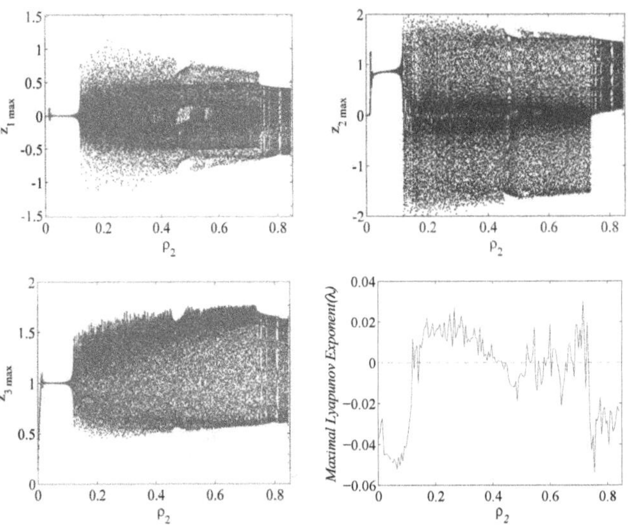

Fig. 7.21: Bifurcation and largest Lyapunov exponents for system (7.2) with variable order $\vartheta(\zeta) = 0.2\sin\left(\frac{1}{0.01+\zeta}\right) + 0.75$.

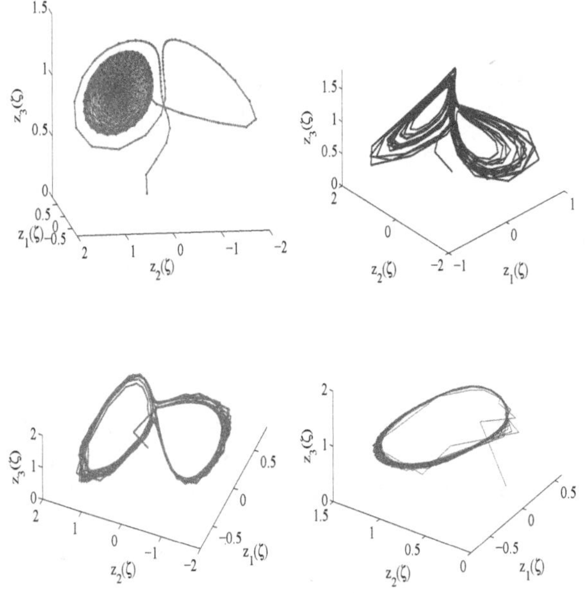

Fig. 7.22: 3-D phase portraits for system (7.19) based on Figure 7.21 with variable order $\vartheta(\zeta) = 0.2\sin\left(\frac{1}{0.01+\zeta}\right) + 0.75$.

7.5.4 Randomness of Variable Order Chaotic Series

Randomness of the chaotic series of incommensurate order plasma model is illustrated with approximate entropy calculated based on (7.9). The randomness analysis is performed for $\rho_1 = 0.6$, $\rho_3 = 1.8$ and

Table 7.6: Approximate entropy of variable order plasma perturbation model.

		$\vartheta(\zeta) = 0.2\sin\left(\frac{1}{0.01+\zeta}\right) + 0.75$		
ρ_1	Tolerance (r)	$\rho_2 = 0.2$	$\rho_2 = 0.3$	$\rho_2 = 0.4$
	$r = 0.2$	0.5247	0.5803	0.5383
	$r = 0.3$	0.4537	0.5019	0.4879
0.5	$r = 0.5$	0.3343	0.3788	0.4020
	$r = 0.7$	0.2402	0.2772	0.3084
	$r = 0.9$	0.1750	0.2004	0.2256
	$r = 0.2$	0.4157	0.5889	0.5913
	$r = 0.3$	0.3592	0.4853	0.5130
0.6	$r = 0.5$	0.2499	0.3183	0.4057
	$r = 0.7$	0.1673	0.2236	0.3059
	$r = 0.9$	0.1236	0.1649	0.2160

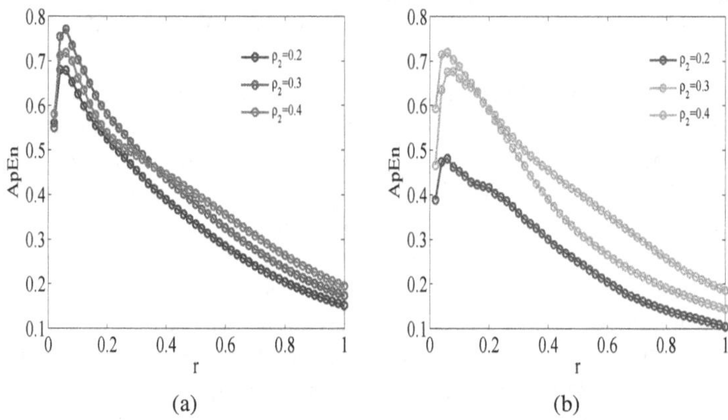

Fig. 7.23: 2-D plot of approximate entropy with varying tolerance (a) $\rho_1 = 0.5$, $W = 1500$, (b) $\rho_1 = 0.6$, $W = 1500$.

$\vartheta = \vartheta(\zeta) = 0.2\sin\left(\frac{1}{0.01+\zeta}\right) + 0.75$ by varying $\rho_2 = 0.2, 0.3, 0.4$. The values are tabulated in Table 7.6 for different tolerance values and lengths of the chaotic series with simulations in Figure 7.23.

7.6 Summary

This chapter focuses on a model of plasma perturbation utilizing a variable order Caputo difference operator. The results obtained from the analysis of chaotic dynamics at two different orders, which are defined as functions of time, are discussed in detail. The findings clearly demonstrate the advantages of incorporating variable orders into a real-world system as opposed to using constant fractional orders. The use of variable orders enables a wider range of chaotic behavior to be observed compared to systems with identical real orders. Furthermore, considering variable orders allows for the inclusion of time-varying characteristics of the state variables, leading to a better understanding of their significance in real-life scenarios. The study also investigates the stability of the system at equilibrium states with both commensurate and incommensurate orders. The regions of chaos and periodic orbits, as depicted by the bifurcation diagrams, are effectively supported by the analysis of the largest Lyapunov exponents.

References

[1] Jan Cermak, Istvan Gyori and Ludek Nechvatal. On explicit stability conditions for a linear fractional difference system. Fractional Calculus and Applied Analysis 18 (2015): 651–672.

[2] Wu, G. C. and D. Baleanu. Discrete fractional logistic map and its chaos. Nonlinear Dyn., 75 (2014): 283–287.

[3] Thabet Abdeljawad. On Riemann and Caputo fractional differences. Computers & Mathematics with Applications 62, no. 3, (2011): 1602–1611.

[4] Adel Ouannas, Amina Aicha Khennaoui, Shaher Momani, Giuseppe Grassi and Viet-Thanh Pham. Chaos and control of a three-dimensional fractional order discrete-time system with no equilibrium and its synchronization. AIP Advances 10, no. 4 (2020): 045310.

[5] Guo-Cheng Wu and Dumitru Baleanu. Jacobian matrix algorithm for Lyapunov exponents of the discrete fractional maps. Communications in Nonlinear Science and Numerical Simulation 22, no. 1-3 (2015): 95–100.

[6] Natiq Hayder, Santo Banerjee, A. P. Misra and M. R. M. Said. Degenerating the butterfly attractor in a plasma perturbation model using nonlinear controllers. Chaos, Solitons & Fractals 122 (2019): 58–68.

[7] Mohd Taib Shatnawi, Noureddine Djenina, Adel Ouannas, Iqbal M. Batiha and Giuseppe Grassi. Novel convenient conditions for the stability of nonlinear incommensurate fractional-order difference systems. Alexandria Engineering Journal 61, no. 2 (2022): 1655–1663.

[8] Liu, Y. Q., A. Kirk, Y. Sun, P. Cahyna, I. T. Chapman, P. Denner, G. Fishpool, A. M. Garofalo, J. R. Harrison and E. Nardon. Toroidal modeling of plasma response and resonant magnetic perturbation field penetration. Plasma Physics and Controlled Fusion 54, no. 12 (2012): 124013.

[9] Constantinescu, D., O. Dumbrajs, V. Igochine, K. Lackner, R. Meyer-Spasche, H. A. U. T. Zohm and ASDEX Upgrade Team. A low-dimensional model system for quasi-periodic plasma perturbations. Physics of Plasmas 18, no. 6 (2011): 062307.

[10] Elsadany, A. A., Amr Elsonbaty and H. N. Agiza. Qualitative dynamical analysis of chaotic plasma perturbations model. Communications in Nonlinear Science and Numerical Simulation 59 (2018): 409–423.

[11] Bothayna S. Kashkari and S. A. El-Tantawy. Homotopy perturbation method for modeling electrostatic structures in collisional plasmas. The European Physical Journal Plus 136, no. 1 (2021): 1–23.

[12] Orain, F., M. Becoulet, G. Dif-Pradalier, G. Huijsmans, S. Pamela, E. Nardon, C. Passeron et al. Non-linear magnetohydrodynamic modeling of plasma response to resonant magnetic perturbations. Physics of Plasmas 20, no. 10 (2013): 102510.

[13] Turnbull, A. D., N. M. Ferraro, V. A. Izzo, Edward A. Lazarus, J. -K. Park, W. A. Cooper, Steven P. Hirshman et al. Comparisons of linear and nonlinear plasma response models for non-axisymmetric perturbations. Physics of Plasmas 20, no. 5 (2013): 056114.

[14] Nikolai Nikolaevich Kalitkin and Dmitrii Pavlovich Kostomarov. Mathematical models of plasma physics. Mat. Model 18, no. 11 (2006): 67–94.

[15] Charles K. Birdsall and A. Bruce Langdon. Plasma Physics Via Computer Simulation. CRC Press (2018).

[16] Diego del-Castillo-Negrete, B. A. Carreras and V. E. Lynch. Nondiffusive transport in plasma turbulence: a fractional diffusion approach. Physical Review Letters 94, no. 6 (2005): 065003.

[17] Devendra Kumar, Jagdev Singh and Dumitru Baleanu. A new analysis for fractional model of regularized long-wave equation arising in ion acoustic plasma waves. Mathematical Methods in the Applied Sciences 40, no. 15 (2017): 5642–5653.

[18] Magdy A. Ezzat. A novel model of fractional thermal and plasma transfer within a non-metallic plate. Smart Structures and Systems, An International Journal 27, no. 1 (2021): 73–87.

[19] Bian, N. H. and P. K. Browning. Particle acceleration in a model of a turbulent reconnecting plasma: a fractional diffusion approach. The Astrophysical Journal 687, no. 2 (2008): L111.

[20] Pincus, Steven M. Approximate entropy as a measure of system complexity. Proceedings of the National Academy of Sciences 88, no. 6 (1991): 2297–2301.

[21] Ahmed Ezzat Matouk. Chaos and bifurcations in a discretized fractional model of quasi-periodic plasma perturbations. International Journal of Nonlinear Sciences and Numerical Simulation 23, no. 7-8 (2022): 1109–1127.

[22] Nasir Ali, Rashid Nawaz, Laiq Zada, Kottakkaran Sooppy Nisar, Zahid Ali, Wasim Jamshed, Syed M. Hussain and Esra Karatas Akgul. Numerical investigation of generalized perturbed Zakharov-Kuznetsov equation of fractional order in dusty plasma. Waves in Random and Complex Media (2022): 1–20.

[23] Amit Goswami, Jagdev Singh and Devendra Kumar. An efficient analytical approach for fractional equal width equations describing hydro-magnetic waves in cold plasma. Physica A: Statistical Mechanics and its Applications 524 (2019): 563–575.

[24] Swarniv Chandra, Sharry Kapoor, Debapriya Nandi, Chinmay Das and Dayita Bhattacharjee. Bifurcation analysis of eaws in degenerate astrophysical plasma: Chaos and multistability. IEEE Transactions on Plasma Science 50, no. 6 (2022): 1495–1507.

[25] Asit Saha and Prasanta Chatterjee. Solitonic, periodic, quasiperiodic and chaotic structures of dust ion acoustic waves in nonextensive dusty plasmas. The European Physical Journal D 69 (2015): 1–8.

[26] Noriyasu Ohno, Masayoshi Tanaka, Akio Komori and Yoshinobu Kawai. Chaotic behavior of current-carrying plasmas in external periodic oscillations. Journal of the Physical Society of Japan 58, no. 1 (1989): 28–31.

[27] Souad Bensid Ahmed, Adel Ouannas, Mohammed Al Horani and Giuseppe Grassi. The discrete fractional variable-order Tinkerbell map: Chaos, 0-1 Test, and Entropy. Mathematics 10, no. 17 (2022): 3173.

[28] Lan-Lan Huang, Ju H. Park, Guo-Cheng Wu and Zhi-Wen Mo. Variable-order fractional discrete-time recurrent neural networks. Journal of Computational and Applied Mathematics 370 (2020): 112633.

[29] Guo-Cheng Wu, Zhen-Guo Deng, Dumitru Baleanu and De-Qiang Zeng. New variable-order fractional chaotic systems for fast image encryption. Chaos: An Interdisciplinary Journal of Nonlinear Science 29, no. 8 (2019): 083103.

Index

F

G

H

I

J

K

L

M

N

www.ingramcontent.com/pod-product-compliance
Lightning Source LLC
LaVergne TN
LVHW081318110826
845149LV00006B/1537

* 9 7 8 1 0 3 2 5 4 4 8 1 6 *